BEYOND THE
SCIENCE WARS

SUNY series in Science, Technology, and Society
Sal Restivo and Jennifer Croissant, editors

Other titles in the series

Peerless Science: Peer Review and U.S. Science Policy, Daryl E. Chubin and Edward J. Hackett

Scientific Knowledge in Controversy: The Social Dynamics of the Fluoridation Debate, Brian Martin

Social Control and Multiple Discovery in Science: The Opiate Receptor Case, Susan E. Cozzens

Human Posture: The Nature of Inquiry, John A. Schumacher

Women in Engineering: Gender, Power, and Workplace Culture, Judith S. Mellwee and J. Gregg Robinson

The Professional Quest for Truth: A Social Theory of Science and Knowledge, Stephan Fuchs

The Value of Convenience: A Genealogy of Technical Culture, Thomas F. Tierney

Math Worlds: Philosophical and Social Studies of Mathematics and Mathematics Education, Sal Restivo, Jean Paul Van Bendegem, and Roland Fischer, eds.

Knowledge without Expertise: On the Status of Scientists, Raphael Sassower

Controversial Science: From Content to Contention, Thomas Brante, Steve Fuller, and William Lynch, eds.

Science, Paradox, and the Moebius Principle: The Evolution of a "Transcultural" Approach to Wholeness, Steven M. Rosen

In Measure, Number, and Weight: Studies in Mathematics and Culture, Jens Høyrup

On the Shoulders of Merchants: Exchange and the Mathematical Conception of Nature in Early Modern Europe, Richard W. Hadden

Ecologies of Knowledge: Work and Politics in Science and Technology, Susan Leigh Star, ed.

Energy Possibilities: Rethinking Alternatives and the Choice-Making Process, Jesse S. Tatum

Science without Myth: On Constructions, Reality, and Social Knowledge, Sergio Sismondo

The Science of Empire: Scientific Knowledge, Civilization, and Colonial Rule in India, Zaheer Baber

In and About the World: Philosophical Studies of Science and Technology, Hans Radder

Social Constructivism as a Philosophy of Mathematics, Paul Ernest

BEYOND THE SCIENCE WARS

The Missing Discourse about Science and Society

EDITED BY
Ullica Segerstråle

STATE UNIVERSITY OF NEW YORK PRESS

Published by
State University of New York Press, Albany

Printed in the United States of America

For information, address State University of New York Press,
State University Plaza, Albany, NY, 12246

Production by Cathleen Collins
Marketing by Anne M. Valentine

Library of Congress Cataloging in Publication Data

Beyond the science wars: the missing discourse about science and society / edited by
Ullica Segerstråle.
p. cm.—(SUNY series in science, technology, and society)
Includes bibliographical references and index.
ISBN 0-7914-4617-4 (alk. paper)—ISBN 0-7914-4618-2 (pbk. : alk. paper)
1. Science—Social aspects. 2. Science and state. I. Segerstråle, Ullica Christina
Olofsdotter. II. Series.
Q175.55 .B49 2000
303.48′3—dc21

99-053390

10 9 8 7 6 5 4 3 2 1

Contents

Preface

The so-called Science Wars would seem like an obvious object of study for practitioners of sociology of science—even more so because this episode directly impinges on the general public's conception of "sociology." Clearly, however, the type of broad analysis that can illuminate this recent episode would have to come from somewhere *outside* the constructivist/relativist approach in this field—not because of the bad name it has recently acquired, but because of its too narrow focus on epistemological issues.

The mid-1990s saw a remarkable campaign by a number of prominent scientists upholding what they called "science and reason" against what they presented as a threat from "the academic (or cultural) left." As so often happens in controversies, the terms of the debate were skewed from the outset. The science warriors felt free to fiercely attack their targets—recent "postmodern" and "constructivist" or "relativist" schools in the humanities and social sciences—in the name of science, while many of their scientific colleagues thought they were going too far. In the same way, the vocal constructivists and relativists saw themselves as representatives for sociology, while their particular approach had been severely criticized by some of their colleagues over the years. There is nothing unusual in this. At the same time, it invites an analysis beyond the positions taken by the contending parties.

Few attempts have been made so far to place the Science Wars in a broader context. *Beyond the Science Wars* strives to do just this. The book applies a multitude of perspectives—historical, sociological, philosophical, scientific, psychological—to situate the Science Wars as a social phenomenon and identify important division lines in this controversy. One of its ambitions is to dispell the idea that the Science Wars represents a clash between the Two Cultures as traditionally understood—a view unfortunately reinforced by the Sokal Hoax. Another is to try to better understand the reasons for the science warriors' persistent talk about "science and reason." What is so special about science for them? That issue is of particular interest also to the authors of this volume, over half of whom have graduate degrees in science.

The book is directed to general readers and academics alike, with a special eye to scientists. The first chapter provides a background to recent developments and an overview of some key events in the Science Wars. Each of the following chapters has an editor's introduction, intended to summarize and contextualize the argument and provide continuity between chapters.

CHAPTER ONE

Science and Science Studies

Enemies or Allies?

ULLICA SEGERSTRÅLE

CONTEXTUALIZING THE "SCIENCE WARS"

This book has three aims. One of them is to try to make sense of the recent debate about science and "antiscience." We will bring light to bear on this question from a variety of perspectives: sociological, historical, philosophical, and scientific. The second aim is to open up a larger discussion about the relationship between the field of science studies and its object, science itself. What is the possible and desirable relationship between scientific practitioners and those who study their activity within science studies, or STS?[1] In the Science Wars, the relationship appeared strained; yet earlier the coexistence between these two scholarly communities could be described as friendly and cooperative. The third aim is to present some missing voices and viewpoints when it comes to the relationship between science and society, including those of two Grand Old Men of STS. Finally, this book hopes to clear up a deep confusion. Unfortunately, the proscience activists in the Science Wars tended to mistakenly collapse the social science–oriented science studies with the new "postmodern" critique of science in the humanities, two quite different academic enterprises. *Beyond the Science Wars* explicitly focuses on STS, although it sometimes broadens the discussion.

This book is directed as much to scientists and the general public as to practitioners and students of STS. The ambition is to clarify a number of issues and raise some new ones that have been suppressed by the very terms of the debate in the Science Wars. Some important things need pointing out from the very beginning. The Science Wars should not be seen as an opposition between scientists and science studies scholars per se. It has been waged by a relatively small *minority* of "proscience activists" against a *particular* school within STS, "the sociology of scientific knowledge," or SSK—a term standing for social constructivist and relativist orientations (and, when it comes to the humanities, against a particular "postmodern" school). Still, since the proscience activists often acted in the *name* of science, and since the social constructivists often were sociologists, the impression may have been given that we here had a deep opposition between science on the one hand and sociology on the other.[2] Not so. The disagreement was between science and the *"social constructivist" and relativist* type of sociology.[3]

Of course, from the proscience activists' point of view, the Science Wars probably started much earlier—some three decades ago with the rise of the social constructivist paradigm in science studies (and in the humanities, with the rise of postmodernism). In this interpretation, it was the social constructivists and postmodern humanists who were the original aggressors, and the 1990s Science Wars was only a reaction by spokesmen for the long-suffering scientists. In any case, it should be noted that during that period, there existed an internal intellectual opposition to social constructivism *within STS* itself, although the voices of these opponents often drowned in the exuberance of social constructivist expansion. This book is not the place to go into the details of this critique, but comprehensive critical analyses exist (e.g., Cole, 1992; Fuchs, 1992; Hagendijk, 1990; Laudan, 1981) and even an alternative program was developed in response to constructivist claims (Schmaus, Segerstråle, & Jesseph 1992).

When initially, in what became known as the "Science Wars," a vocal group of proscience scientists in books, articles, and well-publicized conferences accused some of their fellow academics in the humanities and social sciences for being "antiscience" the attacked parties were taken aback. They felt that it was their academic prerogative to do what they did—that is, treat science as their object of study and subject it to various types of critical analyses, as they saw fit. Moreover, the critical analysis of science had been going on for quite some time without any objections. In the search for explanations for the attack of the "science warriors" (particularly Gross and Levitt with their *Higher Superstition*, published in 1994), it was tempting to try to find the explanation in the existing problems for science, and many did so (Nelkin, 1996 a, b; Ross, 1996). A representative argument was that scientists were in trouble and were scapegoating others; they wanted back to the good old days of abundant science funding.

Indeed, at the time there had been several events of adverse publicity for science, ranging from notorious misconduct cases all the way to the

Unabomber.[4] Some of these cases should probably be considered matters of technology or decision making rather than of science—for instance the Challenger disaster and the problems with the Hubble telescope. Still, these were only some of the problems on the long list included in the special issue "Science under Siege," making the cover of *Time* Magazine in 1992. For many, the most blatant setbacks for science were clearly the closedown of the Superconducting Supercollider in 1993 and the closing of the OTA, the Congress's Office of Technology Assessment. These events seemed to send clear signals from society that the fat years of science spending after World War II and the Cold War were over and that science would be facing serious budget cuts.[5] Still, the adverse situation for science was typically not at all referred to by the proscience warriors themselves at the time. They focused on what they viewed as a threat to science coming from inside academia, from the "antiscience" attitudes that had developed within (parts of) the humanities and social sciences. Let us take a closer look at the academic developments in the sociology of science during the two to three decades preceding the eruption of the Science Wars in 1994.

THE CONSTRUCTIVIST CRITIQUE

Around the mid-1970s, traditional sociology and history of science had given way to new research programs promoting the idea that science was "socially constructed," or suggesting that science was on a par with other knowledge systems, such as Azande witchcraft. These new constructivist or relativist approaches within the newly created field of "the sociology of scientific knowledge" (SSK) postulated among other things that scientific truth had no preferred epistemological status in relation to other truth claims: science was just one among many belief systems, all explainable by social factors. The ground had already been prepared by empirical studies showing that scientists in practice did not follow the "norms of science," that very backbone of the traditional sociology of science promoted by Robert Merton (1942/1973) and his students.

What these new sociologists had dared to do was open up also the content of science to sociological analysis. The Mertonian school had typically treated science as just any other social system, assuming that the norms of science (and their institutionalization in scientific reward and control systems) somehow guaranteed the rationality and objectivity of the knowledge produced. Even the creator of the sister field of "the sociology of knowledge," Karl Mannheim, had declared that natural science was the exception to the rule that knowledge was in general influenced by social ideologies.[6]

One of the great inspirations for this daring move had come from a particular reading of Thomas Kuhn's famous *The Structure of Scientific Revolutions*. For the new sociologists of scientific knowledge, Kuhn was sending a seemingly liberating message: science as an enterprise was not a paragon of rationality after all, it allowed for considerable irrational elements during times of scientific revolutions and scientific paradigm change, during which scientists underwent a type of "conversion" and came to see the world in a totally different way. The new sociologists of science here saw a chance to forcefully introduce *social* instead of philosophical explanations for scientific change—a marvelous opportunity, too, to steal the show from the rationalist philosophers of science who had hitherto dominated the field. With the introduction of the sociology of scientific knowledge, the new sociologists of science had effected nothing less than a self-conscious Kuhnian revolution and paradigm shift within the sociology of science itself! (Kuhn himself was very unhappy, however, with this new "social" interpretation of his thesis, as already an interview I conducted with him in 1982 clearly indicated.)

Many sociologists of science welcomed these fresh new research frameworks proposed by The Strong Programme (promoted by a group at the University of Edinburgh) and The Empirical Program of Relativism (championed by Harry Collins, then at the University of Bath). Others found the French sociologist Bruno Latour's independently developed "actor-network" theory compelling. This Macchiavellian model suggested that science be best studied in terms of a network of actors "enlisting" other actors—and machines—in strategical schemes for winning the scientific game.[7] A whole new academically lucrative research industry now got started, churning out case study after case study, surprisingly supportive of the new dominant paradigm—just as the Mertonian paradigm had earlier supported the vision of science as guided by a particular set of norms.

In all this, the suggestion that sounded most radical to the innocent ear (and to baffled colleagues of the constructivists within STS) was probably the constructivist assertion that scientific *facts* themselves were socially constructed. This view implied that what came to be counted as "facts" was really more a matter of convention or contextual factors than of inherent scientific necessity. The new sociologists of science backed up their claims by ethnographic laboratory studies, which were then cited as exemplary cases by others. (Kuhn had shown how scientists tended to be convinced by "exemplars".) Under such an onslaught of social constructivism, the physical world itself now seemed to come to pieces. Indeed, in the 1980s, many a conference of the Society for the Social Studies of Science was highlighted by ardent disputes between constructivists and realists, or rather, moderates, who dared to insist that the real world did, indeed, constrain science in some ways.[8]

The depth of the constructivist conviction remained unclear even to STS colleagues—when they said that facts were socially constructed, did constructivists mean that there *existed* no scientific facts, or was this merely a rhetorical, or perhaps methodological claim? This seemed never to be satisfactorily resolved even in open disputes. But the important point was not the ontological-sounding claim itself. It was rather that *if* it could be shown that facts could not play the determining role in science that had been earlier attributed to them—for instance, if they could not settle scientific disputes—then the door could be thrown wide open for various types of "social" interests and influences instead; science could be legitimately reduced to a power game.

This also had implications for the history of science. If it could be argued that there "could be" no scientific justification for choosing one theory over another (and there were, indeed, well-known philosophical arguments available), it would seem more legitimate to try to find the social reasons why one theory was historically more successful. The task for the historian would no longer be to describe such things as the working of scientific rationality in a particular historical context, or even the interaction between social and scientific factors—reasonable approaches that did acknowledge the importance also of social factors in science. Instead, the task became to demonstrate the fundamentally social reasons behind even the most abstract-looking scientific ideas. Thus, it was taken for granted that also the scientific convictions of scientists could be unproblematically reduced to social and political interests.[9]

While the new interdisciplinary field of social studies of science was slowly moving into an increasingly constructivist direction, emerging new fields such as cultural studies and women's studies quite independently also chose science as one of their primary objects of analysis. They were interested in studies of the Western bias of science or its inherent masculinity, usually with the implicit or explicit assumption that science as we know it could be otherwise. Students of rhetoric examined the rhetorical strategies of scientists, and the field of literary criticism, inspired by French postmodernism, started treating science as one of many "texts" to be "deconstructed."[10] Also here we had a questioning of the traditional image of science, a downplaying of science as a rational pursuit, and an emphasis instead on science as power. Thus, there was one thing that seemingly united the new science studies and the new postmodernist humanist studies: they both veered away from the idea of science having an epistemologically privileged status.[11]

What about the voice of the scientists themselves, after all the objects of these studies? Unlike earlier sociologists of science, who relied on scientists' own statements, the new science scholars largely ignored what the scientists themselves had to say about their scientific commitments and concerns, or how they judged good science from bad. It is not too much to say that a certain "Besserwisser" approach prevailed, with the sociologists smugly overruling the scientists. It was as if the sociologists were the self-appointed psychoanalysts of scientists, knowing their "true" motives, unbeknownst to the scientists themselves. Unlike earlier sociologists, many of the new sociologists deliberately chose to study scientists using various ethnographic methods. Unlike anthropologists, however, the new explorers did not think that the members of the scientific tribe themselves could act as informants for valuable insights into their world. Meanwhile, for considerable time the objects of study, the scientists themselves, did not seem to be aware of or care about these developments.[12]

1994—THE ANNUS HORRIBILIS AND AFTER

1994 can be characterized as the Annus Horribilis, the year when "the scientists" struck back. But, of course, it was not necessarily "the scientists"—it was rather Paul Gross and Norman Levitt with their book *Higher Superstition: The Academic Left and Its Quarrels with Science*. This book took issue with what it called the academic or "the cultural left," an umbrella term the authors used to bunch together a number of academic endeavors: social constructivists, postmodern humanists, feminists, environmentalists—in short, the many different academic strands engaged in contemporary critical analysis of science. This was also the year of the first of two conferences arranged by The National Association of Scholars, an organization typically working toward bringing back a traditional university curriculum, now almost exclusively engaged in combatting a purported "antiscience" threat. Eminent speakers at these well-publicized events included the Nobel laureate, physicist Steven Weinberg, and Harvard's Edward O. Wilson and Gerald Holton. Already in 1993, Holton set the tone with his essay collection *Science and Anti-Science*, which warned about the dangers of a new irrationalism in society (for a closer analysis of the whole notion of "antiscience," see chapters 4 and 5, this volume).

There were also various indirect skirmishes between representatives for the larger scientific community and its critics, notably in relation to the museum exhibition "Science in American Life." This Smithsonian Institution event was funded by the American Chemical Society and executed by museum curators, who (perhaps with an eye to their academic colleagues) decided to incorporate also social criticism of science in the displays. But the result was soon seen as too negative in spirit and as going against the intent of its original sponsors, the chemists (Flam, 1994; LaFollette, 1996).[13] Incidentally, "Science in American Life" also became the focus of some acrimonious exchange between the two emerging camps in the new debate about science (Gieryn, 1996; Gross, 1996).

As to direct confrontations between scientists and academic science critics, there were at least three memorable ones. One was a famous "showdown" between sociologist Harry Collins and biologist Lewis Wolpert at the British Association for the Advancement of Science (BAAS) in Loughborough, U.K., in September 1994 (see, e.g., Rose, 1996, and Fuller's chapter, this volume). At that conference, it became apparent to many that social constructivists and scientists had difficulty speaking to one another. In any case, the initial (non)exchange of views at BAAS was later followed up in the pages of the *Times Higher Education Supplement* (September 30, 1994). In December the same year, there was a follow-up conference in Durham, U.K. The aim of this second conference was expressly to bring scientists and social scientists together, a move that the organizer, Steve Fuller, himself characterized as "desperate" (personal communication). But that conference did not become the hoped for celebration of mutual understanding. A particular problem seems to have been the focus on case studies, where the two parties could not see eye to eye (for more on the Durham conference, see Fuller, 1995, and this volume).

The third occasion was a panel debate with Gross and Levitt at the Society for Social Studies of Science (4S) Annual Meeting in Charlottesville, Virginia, in October 1995, which I attended. For several reasons, this can rather be described as a nondiscussion. If the aim of this debate was to seriously address the content of Gross and Levitt's book, it was dramatically unsuccessful. Gross and Levitt mostly restated their views, Gross reading a written statement. Among the panelists, the only representative for the many academics who had been criticized in *Higher Superstition* was the feminist Donna Haraway. This part of the debate soon turned into a nasty exchange concerning Haraway's own and other feminist critics' scientific training. What was most disappointing for the ballroom-size expectant audience was that not a single constructivist appeared on the panel or spoke up from the audience. The only good thing was that Gross and Levitt in reponse to questions stated their own irritation with the contemporary "academic left" much more clearly (see chapter 4 this volume).

WHY THE CONSTRUCTIVIST POSITION BOTHERS SCIENTISTS

What happened at more informal meetings between scientists and constructivist academics? One scientist reported that he had attended a seminar where a constructivist asserted that it was in principle possible for there to exist a chemical element between hydrogen and helium in the Periodic Table. For this scientist, this was "strong" constructivism; indeed, he could not see how anyone could seriously hold such an absurd belief (this was a spoken comment in audience at the 4S conference in the session "From Proscience to Antiscience—And Where Next?," which I arranged at the meeting; see also Bauer, chapter 2, this volume). Interestingly, a more common complaint was that "strong" constructivists, when challenged, typically regressed toward the uninteresting and toothless assertion that science is influenced by social factors, that is, a "weak" constructivist stance. It seemed hard to find real, strong constructivists to argue with.

Richard Dawkins (the author of *The Selfish Gene*) appeared to have got lucky, however, since he had, indeed, been able to present a constructivist social scientist with the following question:

> Suppose there is a tribe which believes that the moon is an old calabash tossed just above the treetops. Are you saying that this tribe's belief is just as true as our scientific belief that the moon is a large Earth satellite about a quarter of a million miles away? (Dawkins, 1994, p. 17)

The constructivists' reply was that truth is a social construct and therefore the tribe's view of the moon is just as true as ours. Dawkins now went on to wonder why sociologists or literary critics traveling to conferences did not choose to entrust their travel plans to magic carpets instead of Boeings. "Show me a cultural relativist at 30,000 feet and I will show you a hypocrite," Dawkins concluded (Dawkins, 1994, p. 17).

But what was the rationale behind the constructivist or relativist position? How could the proponents of such views maintain a seeming absurdity of this magnitude? It turns out that the proponents of the new social studies of science were wielding a surprisingly *unsociological* argument as their weapon: an abstract philosophical claim. First they pointed out that science cannot be justified philosophically (they were right—there are, indeed, various problems, most famously the Duhem-Quine thesis, which says that scientific theories are always underdetermined by facts). However, from this abstract reasoning they felt free to conclude that, therefore, in *practice*, too, scientists "could" never have good enough factual evidence to convince themselves and each other, and therefore it "must" be something else that influenced scientific judgment. Social factors! QED. (For this kind of position, see particularly Collins, 1985.)

But one scientist put his foot down in response to this kind of reasoning—and that even before Gross and Levitt. That was Lewis Wolpert, in his book *The Unnatural Nature of Science* (1992/1993). This book appeared to be responding directly to the claims of Collins's "empirical program of relativism," which programmatically refused to grant science any epistemological privileges. (According to that program, the burden of proof was rather on *science* to demonstrate its superiority over common sense

[Collins, 1982, 1985]. We can now see why Wolpert locked horns with Collins at the BAAS in 1994). In his book Wolpert insisted that science *was* indeed different from utility-oriented common sense—science's very wish to understand the world was already "unnatural." Meanwhile, the purported philosophical obstacles for science did not bother Wolpert at all. He simply declared that philosophy of science was of no help for scientists anyway, since scientists had their own criteria for judging scientific theories: such things as parsimony, comprehensiveness, fruitfulness, even elegance. These rules of thumb might not be philosophically justified, but they *worked*, and that was what mattered! (Wolpert 1992/1993).

What probably most upset scientists in general, however—not only the proscience activists—was the suggestion that science was not the objective enterprise it purported to be, or worse, that it could not be objective. This sounded like a combined epistemological and political assault on traditional science, and did indeed appear very similar to the points made by the postmodern and cultural critics of science within the humanities. It may have been on these grounds that Gross and Levitt (1994) classified both postmodern humanism and social constructivism under the common term "cultural constructivism." But while the perceived *effect* of the critical analyses of science may well have appeared to be the same for Gross and Levitt, in reality the nature of the criticisms of these two groups were quite different. What, then, were the differences?

When postmodernist humanists and various types of "standpoint epistemologists," such as a particular brand of feminists, said that science was socially or culturally constructed, they were primarily interested in *values and ideology*. Moreover, the claim was not merely that political, cultural, and personal values *affected* scientific theories—science was seen as *inherently* value laden. For some standpoint feminists, for instance, the very idea of objectivity became a masculine conspiracy (e.g., Harding, 1991).[14] For others, since they saw any theoretical framework as necessarily implying a particular ideological stance (something that could always be demonstrated through a close analysis of a theory's underlying assumptions), even a seemingly theoretical discussion automatically referred to a wider political discourse. Science, therefore, *de facto* became politics, since there existed no objective external arbiter for judging between competing research frameworks or paradigms (e.g., Longino, 1990).

In contrast to this—although Gross and Levitt did not recognize it—the core concern of the sociology of scientific knowledge and its various constructivist and relativist ambitions was not really values and ideology at all. SSK had all the time been primarily interested in *epistemology*, and in demonstrating that a traditional rationalist philosophical explanatory model for science could no longer be justified.[15] The perceived opponents of the social constructivists were in fact the philosophers of science with their rationalist claims. This is why the workers in the new social constructivist paradigm spent enormous energy on showing just how social factors actually enter the knowledge production process, from the involvement of "social interests" (rather than rational judgment) when it came to theory choice, to social (rather than rationalist) explanations for closure of controversies, to social construction or "negotiation" of scientific facts in laboratories, and to the social foundation of all knowledge, including science. (The agenda changed somewhat over the paradigm's lifetime, cf. e.g., Shapin, 1995.)

When it came to values and ideology, however, the sociology of scientific knowledge was in fact sometimes internally criticized for being *insufficiently* concerned with these matters (e.g., Chubin & Restivo, 1983; W. Lynch, 1994, and the discussions in the special issue of *Social Studies of Science*, May 1996, "Politics of SSK"), or uninterested in important social problems of science and technology (see Bauer, chapter 2, this volume). One reason why Gross and Levitt so disliked social constructivism and tended to believe that it was politically motivated, was probably the sharp distinction they themselves made between science and ideology. Science is objective while ideology is socially influenced—or "socially constructed." For them, therefore, *saying that science is socially constructed was the same as saying that science is inherently ideological*—absolutely anathema to their view of science as an objectivist and universalist oasis. This may be why they, in their book, so unproblematically collapsed social constructivism with postmodern humanism, and suggested that "ideology" (or "Theory") were driving both. (For further discussion of science as universalism, see chapter 5.)

Still, it is clear that, independently of its *intent*, social constructivism may well have political *effects*. But what kinds of effects should we assume? Is a book like Harry Collins and Trevor Pinch's recent *The Golem: What Everyone Should Know about Science*, where scientific knowledge is described as fundamentally "contestable," a harmful or positive contribution to the public understanding of science? Fuller in chapter 9 presents these authors as actually wishing to *help* science, by scaling down people's unrealistic expectations of this enterprise. In their book Collins and Pinch themselves, too, appeared to consider it a *democratic* thing to declare that science is "contestable" (they compared it with DNA fingerprinting, which they presented as both unreliable and causing convictions of innocents). People like Gross and Levitt, however, would regard it as a *weakness* if society's progressive forces did not have at their disposal reliable science. For them, the force of the Left is a fundamentally moral one, whose claims about social injustice and inequality rely on potential backup by incontrovertible facts (more on this in chapter 5, this volume).

Indeed, for the many who believed that the scientists who were raising the specter of "antiscience" were social conservatives the best evidence to the contrary might have been the fact that Gross and Levitt among "acceptable" social analysts of science were willing to count Stephen J. Gould, an outspoken left-wing political critic of science. In fact, in Gross and Levitt's book, Gould came off as something of a model social analyst of science! But, if anything, Gould was well-known for his political criticisms of science. How could this be explained? The explanation may be that Gross and Levitt's criterion for an acceptable social study of science was that the researcher should be willing to acknowledge that science was a realm *analytically* separate from politics, even though in practice social values might influence science (Gould's writings often seem to reflect just this kind of position).

Also, it would be incorrect to believe that the scientists on the war path against social constructivism and relativism were uniformly set against all kinds of social studies of science. Indeed, the very same scientists who were opposed to constructivist or postmodern approaches in social studies of science pointed out that they were *not* against philosophical, sociological, and historical studies of science as such. For instance, the physicist Steven Weinberg even declared himself a "history of science buff," and said that he found some types of sociology of science useful and its results plausible. What he disliked was the suggestion that the demonstration of social influences on scientists would affect the *truth* of their theories (Weinberg, 1992, 1996b). Gross and Levitt, too, declared that they approved of traditional philosophy and history of science (Gross & Levitt, 1995; Gross, 1997). (These qualifications became more visible, however, after the impact of their original attack, which could be easily seen as directed against all kinds of historical and social studies of science.)

In a polarized climate, little distinction will be made between, say, constructivist sociologists and nonconstructivist ones—everybody gets tarnished by the same brush, including sociology as a field. And worse, in this kind of conflict the very attempt to *analyze* the situation gets easily misunderstood—by both sides. I have personal experience here, already from two occasions. My sociological analysis of the sociobiology controversy has often been seen by partisans in this debate as either support for sociobiology or for the critics of sociobiology, not as an attempt to understand the debate itself. In other words, the very analysis of a controversy became identified with its object. A vivid example of this phenomenon in regard to the Science Wars was my own attempt to provide a forum for general discussion about the reasons for the seemingly sudden attack on "antiscience" at the Annual Meeting of the Society for Social Studies of Sciences in Charlottesville in 1995. For some reason, the organizers made my session one of the opening ones of the conference (perhaps because it featured sociologist Bernard Barber, who was later to receive the society's Bernal Prize). However, *Academic Questions*, the organ for the National Association of Scholars, later presented my session as but the first of the many constructivist sessions to follow (Zürcher, 1995).

My particular panel ("From Science to Antiscience—And Where Next?") did not feature a single constructivist, something that should have been quite obvious. Indeed, the provocative title of one of the papers, "Antiscience in STS," should have outright pleased the National Association of Scholars' rapporteur. However, in her article, she simply omitted this and other inconvenient data points. She even succeeded in making Grand Old Man, proscience sociologist Bernard Barber seem like a dangerous-sounding feminist! Indeed, all this construction work may well have been necessary, given the title of her article: "Farewell to Reason: A Tale of Two Conferences," which contrasted an "all bad" 4S conference with the "all good" "The Flight from Science and Reason" conference. (Perhaps not by coincidence, the author of this blatant data-selection job later became the Research Director of the National Association of Scholars, I am sorry to report.)

WHO HAS THE RIGHT TO CRITICIZE SCIENCE? A HIDDEN ISSUE IN THE SCIENCE WARS

I now want to move on to one of the fundamental hidden topics in the Science Wars. That is, the question Who has the right to criticize science? Or, put in more technical terms, Who is a competent critic of science? Other scientists in the same field? Any scientist? (Scientists themselves have divergent views on this, see, e.g., Segerstråle 1993.) The public and its representatives? Social scientists, literary critics, and others whose object of study is science? Furthermore, does an academic analyst of science have to master science? (Gross and Levitt kept pointing out that their critics did not know science and, like Wolpert, emphasized the difficult and mathematical nature of science.) What about the public? Does a member of the public have to know science in order to voice criticism?[16]

The proscience warriors seem to have been particularly irritated with the temerity of those who did not have formal credentials in science to criticize their enterprise. From their point of view, what we had here were "outsiders" to science presenting themselves as competent critics of science. How could this be explained at all, except as sheer arrogance? Here it is important to distinguish between the "postmodern" and cultural critique, and the social constructivist or relativist epistemological one. In regard to the former type of science criticism, science was seen as simply part of general culture, ineluctably permeated by social values and ideologies and had no particular basis for its claims to objectivity. In such a situation, it would seem legitimate for "outsiders" to comment on science, because they would no longer be outsiders.

The situation was quite different, however, for the social constructivists and relativists (many of whom, incidentally, did have science degrees). They, too, saw themselves as radical critics of science, but as radicals of an *epistemological* rather than political kind. To quote a spokesman for the "post-Mertonian" sociology of science, sociology of scientific knowledge had seen itself as "new" and "radical," because it was tackling the very *content* of natural science and mathematics, unlike both the "old" sociology of science and traditional Mannheimian sociolology of knowledge. But, this spokesman observed, arguing for the social construction of, say, mathematics, did *not* imply any determined correlation between a *specific* type of mathematics and a *specific* type of social "power arrangement!" It was rather aimed at "counteracting *metaphysical* claims about an 'asocial' foundation for the practices and results of mathematics" (M. Lynch, 1992; my emphasis). Also, Collins in a review article (1983) noted that a typical motivation of the Strong Programmers of the Edinburgh School—the leaders within the new sociology of scientific knowledge—was dissatisfaction with existing paradigms for explaining science, such as Mertonian norms or rationalist philosophy; that is, they were interested in carving out a new *intellectual* niche, rather than in criticizing science per se.

Looking at some of the early publications of that school, however (e.g., Barnes & Shapin, 1979; MacKenzie, 1981), it is hard to consider the Strong Programmers' early concentration on social 'interests' and their connection to social classes as *merely* a sociological analysis without any political connotations whatsoever. For some leading members of the new sociology of scientific knowledge the very act of undermining and "sociologizing" science's claim to a special epistemological status undoubtedly *also* had the connotation of undermining "elitist" scientific expert power in favor of "democratic" common sense knowledge. This was especially true of the Epistemological Program of Relativism (Collins, 1982; Collins & Pinch, 1993). And finally, although the statements by the *originators* of the sociology of scientific knowledge may have represented an apolitical interest in the foundations of knowledge, the *followers* of the SSK program seem to have often regarded their own mission as intellectual-*cum*-political.

Whatever the constructivist intent, for Gross and Levitt and other proscience activists, it was the *consequences* that mattered. And it was is hard to deny that there were potential social consequenses of promulgating a "postmodern" or a constructivist/relativist view about the nature of science. If an impression was created that science as a knowledge system had no particular privileged status in relationship to other ways of knowing, then this could be seen as a de facto legitimation for, say, creationism, or a direct endorsement of "junk science" in the courtroom. In such a climate scientists would have to defend their requests for research money even more fiercely. In practice, therefore, whatever the intellectual content of the Science Wars, it was at the same time a contest about which side, the proscience warriors or the new epistemological critics of science, would be able to define itself as superior in the eyes of society.[17]

We had, then, an interesting situation with two totally different visions of the nature of scientific knowledge pitted against each other. The discussion seemed completely locked. At the same time, it was hard to imagine an external arbiter of some kind—What kind of person could that conceivably be? For the militant scientists, on the one hand, a nonscientist was not competent to speak about science; for the postmodernist humanists and social constructivists, on the other hand, requiring scientific competence might have easily been dismissed as a defensive move or the mystification of expert power. What, in this situation, might conceivably have the power to cut through the Gordian knot of the Science Wars?

THE MEANING OF ALAN SOKAL'S HOAX

One thing that, arguably, did this was the famous Sokal Hoax. The physicist Alan Sokal found a way, despite the impasse, to communicate between the scientific and postmodern worldview, and at the same time conduct what at least for many scientists looked like a crucial experiment. (Note, however, that Sokal's target was the postmodern and cultural criticism of science rather than the social constructivist one within science studies. Still, wide-ranging conclusions for all types of critical analyses of science were typically drawn from the Sokal Hoax.)

In an article in *Social Text,* a leading cultural studies journal, Sokal declared that he as a physicist wished to take the "deep analyses" of certain cultural critics of science one step further by linking them to recent developments in quantum gravity. He wrote about a conceptual revolution with "profound implications for the content of a future postmodern and liberatory science" (Sokal, 1996a, p. 218). For a postmodern piece, his article was suitably entitled "Transgressing the Boundaries: Toward a Transformative Hermeneutics of Quantum Gravity." But in a different journal, *Lingua Franca,* Sokal simultaneously revealed that it had all been a hoax: he just wanted to test whether a piece that was written in the right style and espoused the correct political views, would be accepted as genuine (Sokal, 1996b). And on the face of it, Sokal did prove his point: his article did get published by unsuspecting editors, who were seemingly taken in by his writing. For instance, the following passage was well-tailored to meet the beliefs of the cultural left:

> It has thus become increasingly apparent that physical "reality," no less than social "reality," is at bottom a social and linguistic construct; that scientific "knowledge," far from being objective, reflects and encodes the dominant ideologies and power relations of the culture that produced it; that the truth claims of science are inherently theory-laden and self-referential; and consequently, that the discourse of the scientific community cannot assert a privileged epistemological status with respect to counterhegemonic narratives emanating from dissident or marginalized communities," (Sokal, 1996a, pp. 217–218)

The hoax made the first page of *New York Times* and triggered an avalanche of discussion in that paper and elsewhere. But what were the real implications of Sokal's hoax? For some it suggested that editors of cultural magazines were not capable of distinguishing serious from nonserious reasoning as long as the *form* was right and the article reached expected political conclusions. For others, it indicated that the editors may have been pleased by an apparently postmodern contribution coming from a scientist. Still for others, it suggested that Sokal's political qualifications as a leftist (he represented himself as having worked in Nicaragua during the Sandinista regime) had misled the editors about his true convictions (that he was not a cultural leftist). Indeed, in their response, the editors wrote about deception and break of trust—a surprisingly nonpostmodern complaint (Ross 1996). (See also Stanley Fish, 1996, speaking for cultural studies.)

Now Sokal's hoax may be used as a just-so story for many things. For some, it seemed like the ultimate test of the possibility of communication between the Two Cultures, while it highlighted an important asymmetry in each culture's capability to assert the academic power of its own position. Did the success of the hoax perhaps suggest that physicists could figure out what it took to get published by humanists, but not vice versa? Or did the result demonstrate a point often raised in the Science Wars, that those who passed critical judgment on science did not know its content—in this case, not even what passed for a reasonable scientific argument? (Sokal had made sure that his scientific documentation was impeccable, but he had made deliberate "funny" mistakes in scientific inferences, not clearly perceptible to nonexperts.) If so, in both cases, Sokal would in fact have supported the received view that science *was* more "difficult" than the humanities—a view already conveyed by Gross and Levitt.

Sokal himself affirmed that he had got inspired exactly by reading Gross and Levitt. As he explained at the conference "Science and Its Critics" at the University of Kansas in late February 1997, when reading *Higher Superstition*, he had wondered if the quotes of postmodernists were really representative or if they had perhaps been taken out of context. So he had gone to the library to find out, he said—and lo and behold, the quoted passages were even worse in context! Sokal assured the audience that his hoax was "really very, very funny," but noted that the most hilarious part of his article was not even written by him—he had simply quoted the silliest quotations he could find. "Don't miss my footnotes!" he said and giggled to the audience.

In the same talk, Sokal also professed his surprise at the fact that his prank had reached the front page of the *New York Times*, and that "his name has now become a verb." According to Sokal, the whole thing had taken on a magnitude ten times bigger than expected. He also said that he disliked the term "Science Wars." Still, his hoax was undeniably fed into an ongoing debate, and young Sokal could hardly have failed to realize that he had now de facto aligned himself with the science warrior camp.

Obviously, too, the aim of his article was not only to poke fun at postmodern jargon. Sokal's *Lingua Franca* revelation shows an attitude quite similar to that of Gross and Levitt (who were, indeed, very pleased with him; Gross & Levitt, 1996). Sokal described his hoax as an "experiment" to test the intellectual standards of a certain academic subculture. According to him, the results "demonstrate, at the very least, that some fashionable sectors of the American academic Left have been getting intellectually lazy." But why did he choose the medium of parody? Why not simply demonstrate to the postmodernists that they were wrong? Sokal explained that the subculture "typically ignores (or disdains) reasoned criticism from the outside," so parody was the only way to get through to them:

> In such a situation, a more direct demonstration of the subculture's intellectual standards was required. But how can one show that the emperor has no clothes? Satire is by far the best weapon; and the blow that can't be brushed off is the one that's self-inflicted. I offered the *Social Text* editors an opportunity to demonstrate their intellectual rigor. Did they meet the test? I don't think so. (Sokal, 1996b)

Through his action, however, Sokal did not only hoax the editors of *Social Text*. What is less known is that he in this way subverted the radical intent of that whole issue of *Social Text*. That particular issue entitled 'The Science Wars' was in fact exactly intended as a cultural left response to Gross and Levitt (Ross, 1996a). In that issue, the Sokal piece ended up somewhat tagged onto the other pieces, which were all dealing with the recent rift between science and cultural studies of science. (Sokal's piece was duly omitted in the later book version of the special issue; Ross, 1996b.)

And there were further implications. The hoax soon became a convenient vehicle for those wishing to prove points about cultural studies of science. Indeed, Sokal himself in his early response had epitomized the "hardline" scientific attitude in regard to cultural studies. He stated categorically: "*Social Text*'s acceptance of my article exemplifies the intellectual arrogance of Theory—that is postmodern literary theory—carried to its logical extreme," where "[i]ncomprehensibility becomes a virtue; allusions, metaphors and puns substitute for evidence and logic" (Sokal, 1996b, p. 63). Against this he asserted his own position:

> There *is* a real world; its properties are *not* merely social constructs; facts and evidence *do* matter. What sane person would contend otherwise? And yet, much contemporary academic theorizing consists precisely of attempts to blur these obvious truths. (Sokal, 1996b, p. 63)

It was only later that Sokal started backing off from the intent and implications of his hoax. For instance, he began his talk at the Kansas conference by pointing out that "not much" could be deduced from the fact that his hoax was published. He said it did not prove, for instance, that intellectual standards were lax in general, and so on, "only that *one* journal published an article that they *admitted* that they could not understand, solely because it came from a 'credentialed' person." This was obviously a sound conclusion by a scientist who had, after all, only one data point!

This raises an interesting question when it comes to the proscience activists' attitude to Sokal's hoax. To be consistent, should not Gross and Levitt have treated the Sokal hoax in the same way that they treated instances of scientific fraud—declaring it an isolated case, proving nothing? Gross for example, downplayed scientific fraud as "based upon no frequency data, absolute or relative to other professions" and as merely "a few, highly publicized misconduct cases, some of which have been dismissed" (Gross, 1997). But when it came to Sokal's hoax, the interpretation was quite different. The hoax was not regarded as a lonely data point, calling for real frequency data. Instead, it was seen as a type of *legal precedent*. In the Sokal case, hard-line scientists had made a surprising move from their usual quantitative scientific standards toward a case-oriented legalistic attitude! This was reflected, for instance, in Gross's own contention that the targets of the Sokal hoax ought to have learned a lesson from the hoax:

> You might imagine that with egg on their faces, cultural studies entrepeneurs would have repaired to the ladies' and gents', respectively, to wash up, resolving to do better or at least to attend henceforth to the content of the science they study. But no. Academic scholarship used to be like that; now it is not. Some parts of academic life are a political game—as the best players insist—like all other human activities. So the spokespersons for STS deal not with the arguments of the opposition but with its motives. (Gross, 1997)

(Note here the surprising slide from "cultural studies" to "STS," a move that we will return to later. Although the Sokal hoax was making particular fun of the way in which postmodernists were treating science and had nothing to say about social studies of science, his hoax was generally seen as proving a more general point in the Science Wars.)

Whatever Sokal did or did not intend, the deed was done. After this, the Sokal hoax took on a life of its own, while its perpetrator was off to new pastures. After an extensive academic lecture circuit Sokal began working on a critique of the way French postmodernists (mis)use physics together with the Belgian physicist Jean Bricmont, resulting in the book *Impostures Intellectuelles* (Sokal & Bricmont, 1997).[18] Meanwhile, the debate about the meaning of the hoax continued. What Sokal did not know was that his hoax would later become the subject of an article by Steven Weinberg in the *New York Review of Books* (Weinberg, 1996a), which in turn would create its own reaction.

WHO OWNS THE HISTORY OF SCIENCE? THE AFTERMATH OF THE SOKAL HOAX

Weinberg's article was ostensibly devoted to bringing out the really funny parts of Sokal's hoax, lest the layman, ignorant of modern physics, would miss out on the joke. In his article, however, Weinberg did not only explicate some fine points of physics but also clearly asserted the difference between what he called the "inner logic" of science and (what he took to be) the social constructivist position. This article drew several responses. Two Yale professors, together teaching a course on literature and science—one from the viewpoint of comparative literature, the other from the perspective of science—accused Weinberg of assuming a mantle of purity, while boiling down "science" to the work done by particle physicists (Holquist & Shulman, 1996). They objected that outside Weinberg's "reductionist temple" would be found not only sociologists, historians, philosophers and postmodern theorists, but also many famous physicists and whole scientific fields. They noted that Weinberg's "obsessive dualism," where the timeless laws of science were posited against the social world of culture, reminded them of the dualism separating the profane from the sacred.

In the same issue the director of the Center for the Critical Analysis of Contemporary Culture at Rutgers, George Levine, protested against Weinberg's contention that the conclusions of physics could have no cultural implications. He called this "an extraordinary, a profoundly irrational claim" and retorted that it was hard for laypeople *not* to draw cultural inferences from, say, Bohr's complementarity principle. The gist of Levine's critique was that the counterattackers of postmodernism and science studies had themselves become irrational and unscientific in the very name of science (Levine, 1996).

A similar thought was echoed by the Princeton historian of science, Norton Wise (Wise, 1996). Wise accused Weinberg of presenting "an ideology of science, an ideology which radically separates science from culture, scientists from 'others' and splits the personalities of scientists into rational and irrational components." According to Wise, the history of physics, particularly quantum mechanics, was full of scientists who had been motivated by philosophical, political, and other beliefs. He ended by asking whether Weinberg was trying to promote a cultural agenda of his own, attempting to rewrite history.

This remark could have cost Wise the position of special Research Professor in Science Studies at Princeton's Institute of Advanced Studies in May of 1997. At least, this is what was suggested by a report in the *Chronicle of Higher Education* (McMillen, 1997). The article theorized that a campaign by active proscience warriors succeeded in blocking Wise's appointment, just in the same way it had done years before, when Bruno Latour had been a candidate for the same position. It further suggested that Wise's recent polemics with Weinberg might have played an important part in this matter (McMillen, 1997). The outcome was that the position of Research Professor in Science Studies was not filled at all—just as had happened in the Latour case.

Pursuing this line, we find that Wise had in fact a longer record of "offenses," such as a sharp and much noted review of *Higher Superstition* in *Isis* (Wise, 1996). And he was, of course, the Chair of the Department of History of Science at Princeton, in turn directly associated with Gerald Geison's controversial *The Private Science of Louis Pasteur* (Geison, 1995), a book which in 1996 produced a sideshow to the Science Wars in the pages of the *New York Review of Books*. There molecular biologist Max Perutz criticized the book severely as "bad" history of science, suggesting among other things that Geison made too much of Pasteur's supposedly unethical behavior because he did not know chemistry (Perutz, 1996). This critique in turn led to a heated interchange between Perutz and supporters of Geison (Summers, 1997; Perutz, 1997).

Meanwhile, Weinberg himself in his response to his critics (Weinberg, 1996b) made the whole situation largely sound like a misunderstanding. He declared that he had no quarrel with most historians, philosophers, and sociologists of science, but was, rather, concerned with "the corruption of history and sociology by postmodern and constructivist ideologies." He said he accepted the idea that people could be *inspired* by scientific metaphors, but not the idea that science had any clear cultural *implications*. He also pointed out that he was not speaking for science in general, only for physics. In an important passage, however, Weinberg now also pinpointed what he himself believed to be the fundamental difference between himself and Wise and other letter writers. The latter's "agenda," according to Weinberg, was "to emphasize the connections between scientific discoveries and their cultural context." Weinberg agreed that scientists might well draw inspiration from cultural influences, but cultural influences later got sifted out. They did not become a permanent part of scientific theories:

> Whatever cultural influences went into the discovery of Maxwell's equations and other laws of nature have been refined away, like slag from ore. Maxwell's equations are now understood in the same way by everyone with a valid comprehension of electricity and magnetism. The cultural backgrounds of the scientists who discovered such theories have thus become irrelevant to the lessons that we should draw from the theories. (Weinberg, 1996b)

The issue for Weinberg, then, was "not the belief in objective reality itself, but *the belief in the reality of the laws of nature*" (italics added). For Weinberg, that specifically meant the lack of "multiplicity," that is, the existence of different laws for different cultures. At the same time, he saw the belief in multiplicity as a logical consequence of a cultural contextualist stance.

The polemics around Weinberg's Sokal article and its aftermath nicely clarified many of the issues that divided the two camps in the Science Wars. We see that we are not only dealing with an opposition between a constructivist/relativist and a realist outlook on science and the world. Weinberg's statements could be interpreted so that he (together with some other hard-liners) would wish to move in on the very territory of the humanists! *There seemed to be an open struggle between the two camps as to who "owned" the history of science!* And the fact that opposing Wise's Princeton appointment were not only scientists but also historians and historians of science (McMillen, 1997) suggested that the dividing lines went deep indeed in the Science Wars, also in regard to the history of science. (There was later a countermove by historian of physics Silvan

Schweber, who, with address to Steven Weinberg, suggested that the laws of physics might not, after all, be written in stone [Schweber 1997]. And in 1998 Mara Beller suggested that some of the blame for the excesses of the postmodernist critique of science ought to fall on physicists such as Bohr, Born, Heisenberg and Pauli, whose "astonishing" philosophical pronouncements were "hardly distinguishable from those satirized by Sokal," p. 30).

In other words, the proscience warriors appeared to be greatly concerned with protecting a particular understanding of science—that is, the "true" one—and deeply resented the fact that postmodernists, social constructivists, and even historians of science were meddling with science and representing it in other ways that were "false." They did not seem to recognize the right of other academics to do their own interpretations of science within the particular frameworks of their own disciplines. (One problem was, of course, that those interpretations were never really kept within the disciplines—on the contrary, they often became part of undergraduate curricula, or publications aimed at the public understanding of science. And it was clearly this that mattered to the proscience fighters.)

THE SCIENCE WARS CONTINUE: CLOSING IN ON SCIENCE STUDIES

The humanists on the one hand and the science studies community on the other may have been initially taken by surprise, but they did finally react. A quick cultural studies response was to publish an expanded book version of the "Science Wars" issue of *Social Text*—sans Sokal (Ross, 1996b). The STS community's initial response, again, was a book review in its leading journal *Social Studies of Science*, pointedly entitled "Social Construction of an Attack on Science" (Martin, 1996). Later, a new round of reactions was triggered by Gross, Levitt and Lewis's edited book *The Flight from Science and Reason* (1996), based on the New York Academy conference with the same name. In this round it was particularly Paul Forman's critical review in *Science* (Forman 1997a) that caused angry protests by letter writers, some of which were scientists who had actually contributed to the volume. They accused Forman, a historian of science, for "postmodern" scholarship and for completely misrepresenting the content and spirit of the book (Hershbach, 1997; Levitt, 1997; Robinson, 1997; Trefil, 1997). Forman remained unperturbed (Forman, 1997b). Still, there were suggestions that the angry responses to this and other science-skeptic contributions may have contributed to the resignation of the book editor of *Science*.

Later, there were signs of escalation of the warfare from the militant scientists' side. In the late spring of 1997 Gross suddenly decided to abandon his attack on "postmodernism" or "the academic left" and go after

STS instead (Gross, 1997). In an Opinion paper in *The Scientist*, a biweekly journal directed to practicing scientists and science administrators, Gross dismissed or mocked the views of leading STSers that had appeared in an earlier interview article in that journal. He also here took the opportunity to rebut Dorothy Nelkin's (1996b) suggestion that the science warriors were looking back to the good old days of post–World War II funding. But most surprisingly, *Gross now presented STS as a powerful "international movement," amply funded by governments!*

Thus, in April 1997, the threat was no longer "the academic left"—it was instead "STS," which was now made to sound almost like a Communist conspiracy! Note, too, that Gross now refused to make any distinction between STS and such things as Afrocentrism, feminism, and antiwesternism. On the contrary, he explictly described the program of "some of the most influential STS practitioners" as being to belittle science and scientists and as "part of the vogue, among the intelligentsia, for derogation of Western culture." Gross even presented antiwesternism as central to the "business" of STS:

> Within important precincts of STS, science is just another business; the STS job, as more than one admired practitioner has explained, is to put science in its place among the other belief systems. It insists that science is just another narrative that is beholden to, constructed by and for white, European, capitalist patriarchy. (Gross, 1997)[19]

Gross further warned his readers that the authors of some "iconic works of science" "hope for high positions in scientific policy-making." (Whom did he have in mind? Latour?) But, Gross continued, STS is not needed! Science can take care of itself. He went on to say that STS claims to have discovered failures of scientific self-regulation, but there is no evidence of widespread ethical failure. Science is not irresponsible, and so does not need STS to set it straight! STS is not needed even for pointing out social abuse of science and technology; also in these matters, it is typically the scientists themselves that have taken the lead, Gross asserted (Gross, 1997). So much, then, for STS.[20]

How might we explain this sudden attack on STS, and the identification of STS with antiwesternism, of all things? Already in *Higher Superstition*, Gross and Levitt seem to have regarded the various types of cultural criticism of science as an offshoot of the Culture Wars from the 1980s. They made this connection at the beginning of their book, where they said: "Natural science is one of the last major features of Western life and thought to come systematically under the critical gaze of the academic left" (p. 5). Note, however, that in *Higher Superstition*, they did seem to recognize the internal differences between the different strands of radical thought that they were criticizing. This was indicated by the fact that they devoted different chapters to such things as postmodernism, feminism, and science studies, and then used the 'academic' or 'cultural' left as an umbrella term for all of these together. It is hard to believe that at this stage of the Science Wars, the sudden introduction of "STS" as a new global term was a simple mistake.

So what could conceivably have been the point of using the acronym STS in a maximally negative way in an article directed to scientists (and science administrators) at the end of April 1997? The timing suggests that this particular attack on STS *could* have been related to the blocking of Norton Wise's appointment to the research professorship in Science Studies at Princeton—perhaps it was part of a "climate creation" effort often seen in politics.[21]

In the meantime, the science studies community had started developing various defensive strategies. David Edge, the editor of the community's core journal *Social Studies of Science,* soon assumed a leading role, devoting special editorials and postscripts to record keeping of any and all incorrect representations of social constructivism and STS by people like Gross, Levitt, and Weinberg. Edge also informed his readers about ongoing polemics in *Nature* and *Science,* in which he himself took active part, and published in full "for the record" the contents of his own rejected letters to these journals. At one point the STS community was even said to have considered lawsuits, but realized that these were impractical or impossible in a trans-Atlantic setting.[22] Another defense strategy was to describe the position of the Strong Programmers (and by implication, also that of the rest of the sociologists of scientific knowledge, SSK) as part of *science,* and the practitioners in this field as meticulous, good *scientists* (Edge, 1996a–d; 1997a–c,; Barnes, Bloor, & Henry, 1996; Bloor, 1997). According to both Edge and the Strong Programmers themselves, they had always been acting as scientists.

The British kind of counterstrategy against proscience writers, again, has typically directed itself toward the heart of the matter in that country: the public understanding of science. (See Fuller, chapter 9, for the role of the public understanding of science in Britain.) The strongest argument in this genre was undoubtedly *Misunderstanding Science*? (Irwin & Wynne, 1995), which questioned the taken-for-granted assumption that the public would love science more if they only knew it better (the basic argument, for instance, in a certain Royal Society report). Not so, said the authors: the public did understand science, it saw its social impact—and it was exactly because of this that it was skeptical!

We can note a certain symmetry in the approaches used by both the science activists and STS scholars in the Science Wars. For instance, in a kind of inversion strategy, at one point *both* sides played victim, assuming the role of a beleaguered minority. Dorothy Nelkin started it by suggesting that "[a] surprising number of scientists are attacking the works of social scientists and humanists" (Nelkin, 1996b). Gross retorted that the number of members in "the anti-antiscience band" (his term) was only "less than a dozen." STS, in contrast, was a well-funded and forceful international movement (Gross, 1997). Moreover, both sides reached out to their constituencies. Thus, Gross warned the scientific community that they had better start reading the literature to see for themselves how science was being currently misrepresented, while Edge protested in editorials in *Social Studies of Science* about the tone and content of certain remarks by Gross, Levitt, and Weinberg (Edge, 1997a–c). And neither side failed to use the presumably ultimate weapon in the Science Wars: accusing the other side of being "unscientific." Gross and Levitt had, of course, started this, by making much of their opponents' supposed ignorance of science. But Edge now served Gross and Levitt some of their own medicine, charging them of not living up to scientific standards themselves in their book. (It should, though, be noted that Gross and Levitt themselves saw their attempt as polemical.)

An unexpected, positive outcome of the Science Wars, however, was that also "regular" practicing scientists (rather than proscience warriors) started becoming interested in the debate. These scientists took the initiative to several new conferences and discussion fora, with the aim of establishing open communication between representatives for the Two Cultures.[23] "Regular" (not activist) scientists also started discussing matters in *Nature* and especially *Physics Today* (Gottfried & Wilson, 1997; Gottfried, 1997; Mermin, 1996a,b, 1997). They seem to have been particularly interested in the claims of constructivists and the case studies in Collins and Pinch's *The Golem* (Collins & Pinch, 1993). And to this Collins and Pinch responded (Collins & Pinch, 1996, 1997), although they did not move from their position.

The discussion triggered by the Science Wars seems to have further stimulated the interest among physicists particularly. A recent example was an invited symposium "Science and Its Critics" at the 1998 annual meeting of the American Physical Society which featured Kurt Gottfried (physics, Cornell), George Levine (humanities, Rutgers), and myself. The *Newletter* editor commented: "The symposium was notable for the mutual respect reflected in the exchanges" (*History of Physics Newsletter 7* (3), Fall 1998, p. 7).

But just as in other academic debates, the issues that were debated in the Science Wars were not necessarily the ones that were most important. One "hidden issue" was the relationship between science and society. The Science Wars at least in part reflected the lack of clarity in science's basic social contract at the end of the twentieth century. As many have observed, the postwar growth curve of American science had tapered off, giving way to a new type of "postacademic" science, with new sponsors, new rules, and new criteria for "good science" (e.g., Ziman, chapter 7 in the present volume, Gibbons et al. 1994; Slaughter & Leslie, 1997; Stephan, 1996). In other words, the conditions informing the basic assumptions of the post–World War II contract between science and society had changed, which meant that the relationship between science and society may have to be completely rethought (Byerly & Pielke, 1995). Within this broader picture, the Science Wars emerge somewhat like continents on top of the underlying plate tectonics of academic science, industry and government, all shuffling against one another.

From the point of view of what really needs discussing, the Science Wars appears as a rather odd phenomenon, circling around esoteric issues rather than addressing real ones. Ironically, the specialists when it comes to matters of science and society are (or ought to be) the science studies scholars—at least in principle. It is time for the science studies community to resume responsible intellectual leadership in matters having to do with the relationship between science and society, in cooperation with our scientific colleagues. This would also mean a reorientation of the field of STS toward it original mission, including the question of science and social values and the moral responsibility of scientists—a problem that was pushed under the carpet first by the epistemological trend in the science studies and later by the focus of the debate in the Science Wars (cf. Bauer, chapter 2, and Fuller, chapter 9).[24]

PLAN OF THE BOOK

This book attempts to understand the various sources of the recent tension between vocal subcommunities within science and science studies (STS). It also strives to place the Science Wars in a broader context. We need to know how this episode relates to a larger historical record of pro and antiscience sentiments, earlier conflicts between the Two Cultures (if this was, indeed, a Two Cultures battle), and comparable recent episodes where scientists have cried "antiscience!" Why were the proscience activists so emotional about cultural and social studies of science and what exactly was it that they objected to? Moreover, did they correctly understand the claims of those whom they criticized? And what did they themselves mean with "science and reason"? Meanwhile, where is science itself going, and how does this fit in with the scientific worldview of the proscience warriors? Finally, looking at the historically changing relationship between science and social studies of science, what is the possible and desirable relationship between these two groups of academics?

We begin with a protest from a scientist, who has also for a long time been professionally engaged in science studies and is the author of a number of books in this field. In chapter 2 Henry Bauer believes current science studies have gone too far. He sees in them an antiscience attitude that totally diverges from scientists' own conceptions of science. He admits to a conflict between his taste as a scientist and the interpretive and philosophical orientation of much of current science studies, but he also registers clear irritation with "absurd" claims. He recommends that social studies of science abandon their current obsession with epistemology and return to *social* issues related to science and technology—also for strategical reasons. (As might be expected, the reader will perceive a certain tension in the text between Bauer-the-irritated-scientist and Bauer-the-reformer-of-STS).

We should not forget that, until recently, there were good relations between science and science studies, or rather, the fields of history, sociology, and philosophy of science. In chapter 3 Bernard Barber, who helped found the sociology of science after the Second World War, shares with the reader his own sense of the changes taking place in this field over half a century. We learn the intriguing fact that his 1952 book *Science and the Social Order*, often read as an argument for the close connection between science, "Mertonian norms" and democracy, was appreciated both by British Marxists and the CIA. (Of course, after Sputnik in 1957, the book required a new reading.) Meanwhile, whether he intended it or not, Barber himself helped instill a critical attitude in many with his famous 1961 *Science* article "Resistance by Scientists to Scientific Discovery." (Incidentally, it was this article that inspired at least two contributors to this book to go into social studies of science.) Barber's verdict on recent developments in the field is that innovations like the Strong Programme have had a positive impact, but now need to be transcended, for progress to be made.

Chapters 4 and 5 confront head-on the many recent claims about "antiscience," while also trying to understand the motivations of the anti-antiscience warriors. In her search for the meaning of "antiscience," the author, a sociologist of science, compares recent allegations to two episodes in the 1970s and '80s where scientists also cried "antiscience!": the American conflict about behavior genetics and sociobiology, and a case in Britain, where two physicists blamed philosopher Karl Popper for the decrease in science funding. She ends up questioning the existence of a clear antiscience category, and argues that the label "antiscience" might be seen as a *heuristic device* that allows academics from different fields to collectively combat various critics of science—sometimes including their own scientific opponents as well! The author further argues that an important motive for the anti-antiscience warriors was the idea of science as an universalist and democratic *cultural* force. However, they too narrowly identified "reason" with *scientific* rationality, which contributed to the polarization in the Science Wars.

What about the broader context of the alleged antiscience phenomenon? Under what circumstances have attitudes to science typically been positive or negative? In chapter 6, Valéry Cholakov, a historian, gives examples of the changing social support for science in the 20th century. He discusses three metaphors depicting science, "science, the endless frontier," "science, a direct productive force," and "science, the land of opportunity." These slogans were tied not only to national security interests in the United States, the former Soviet Union, and Germany, respectively, but also to science's earlier dramatic and visible contributions, which made it possible to present science as a positive social force and problem solver. The change from the post–World War II proscience attitude to the 1980s' skepticism about science is in part a reaction to science's oversell and the fact that it is not any longer bringing about quick changes in everyday life.

The following chapter takes a futuristic look. John Ziman, a theoretical physicist turned theorist and statesman of science, discusses how the change in scientific patronage within "postacademic science" will affect the structure and content of science as we know it. Noting the imminent changes in the organization of scientific research predicted by recent books—a move toward more local and applied practices and norms—Ziman worries that abandoning traditional academic science may present a danger to *democracy* as we know it. He sees academic ("pure") science with its universalist and objectivist ambition as intimately tied to democratic ideals, and as therefore representing an important cultural model for fact-based societal dialogue. We abandon academic science at our peril, warns Ziman. As we see, unlike Gross and Levitt, Ziman presents the threat to science as the result of a process of *structural change in science itself*, rather than as caused by "antiscience" sentiments.

The issue of perhaps greatest concern for the science warriors is the postmodern challenge to an objectivist epistemology. In chapter 8, Stephan Fuchs, a sociologist, reviews the several recent efforts by historians and sociologists of science to "locate" objectivity in various ways, typically emphasizing such things as particular communication practices, or trust. Fuchs's target is not these academics, but rather some of the most vehement feminist "standpoint epistemologists," who question the very value of objectivity. After first refuting their claims, he suggests that we abandon epistemological musings altogether and get back to a *sociological* analysis. With the sociologist Niklas Luhmann, he argues that "science" represents a particular mode of socially indispensable communication exclusively oriented toward questions of truth versus falsity and to the elimination of error. In other words, also Fuchs emphasizes the social function and intent of science. We can now begin to understand one mystery in the Science Wars: how it was possible for proscience scientists and other academics to so ardently want to defend science, even if they had rather different views of what science was. The one thing they all seemed to agree on, however, was that science was a socially important institution because of its ability (or at least intent) to separate true claims from false.

Finally, in chapter 9 Steve Fuller, one of the most active representative of the STS community in both Britain and the United States, reflects on some of the ironies of the Science Wars. He traces the relationship between science and science studies all the way back to August Comte and forward to Thomas Kuhn, stopping on the way to comment on the (surprising) meaning of C. P. Snow's *Two Cultures* in that book's historical context. According to Fuller, the Science Wars is *not* a Two Cultures war, already because only a fraction of each "culture" is involved. He notes further the irony that the STS program of the Edinburgh School in fact originated in a post–World War II campaign to "humanize" scientists initiated by scientists themselves. However, the leaders of this effort later started playing the academic game and turned to more prestigious theoretical pursuits in the sociology of science. Fuller also demonstrates, that *contra* Gross and Levitt, there is no historical connection between STS and postmodern humanism. We get some tales from the front, among others from the Durham conference which Fuller arranged to generate dialogue in the Science Wars, and a refreshing analysis of the Sokal hoax. Finally, Fuller discusses the problems and possibilities of a reconciliation between scientists and science studies scholars.

BRINGING THE SCIENTIST BACK IN

The aim of this introduction has been to give the reader an overview of some important events and actors in what has come to be known as the Science Wars, and suggest some larger frameworks that can be used for understanding this debate. I have also tried to indicate that the present seeming impasse between science and science studies is a contingent, not a logically necessary one. There have indeed been times when scientists on the one hand, and sociologists, historians, and philosophers of science on the other, have been living in perfect harmony; what is more, cooperated, and there is no reason why this would not be possible again. (On the face of it, cooperation would seem like the natural state of affairs if we are interested in the nature of scientific knowledge.) Moreover, it is important to remember that the parties in the

Science Wars—proscience scientists and constructivist/relativist sociologists of scientific knowledge—only represent *particular*, although very vocal, subcommunities within the larger scientific and STS communities respectively.[25] There is a need for other voices to be heard. *Beyond the Science Wars* is a deliberate attempt to bring nonconstructivist views in STS to a larger audience and provide a set of perspectives for a broader discussion and dialogue.

The problem with the recent developments in the sociology of scientific knowledge has not been that its proponents do not know science, but rather that many practitioners in this field have not seen the need for a dialogue with their objects of study. (In fact, some have even regarded it as important *not* to listen to scientists, or as an advantage not to know science.) This is why this book is at the same time a call for a new sociology of science that will aim at "bringing the scientist back in" (cf. Segerstråle, 1989, 1993). The question is *how to build a sociology of scientific knowledge—and science—that can also be recognized as true and useful by practicing scientists themselves*.

One such idea is "The Hard Program in the Sociology of Scientific Knowledge: A Manifesto" (Schmaus, Segerstråle, & Jesseph, 1992). The Hard Program is "hard" among other things because it eschews sociological reductionism and argues instead for a type of comprehensive sociological explanation of science that also considers factors *scientists themselves* consider important, such as various types of cognitive commitments. Moreover, it retains a "reality check," by inviting scientists, too, to comment on the research results, and even a critical function, since it allows for particular cases of scientific behavior to be compared to prevailing conceptions of "good science" at the time. Needless to say, this is only one of many possible avenues; many more will probably appear in the wake of the Science Wars.

But bringing the scientist back in can mean even more. In fact, the best defense for STS and the sociology of science may be the growing interest of "regular" scientists in the actual workings of science. In the future, the real challenge to science studies scholars may not come from proscience activists, but rather from practicing bench scientists enthusiastically scrutinizing the STSers' findings. Perhaps with the help of our new collaborators, the scientists, who as a result of the Science Wars have been stimulated to introspect and reflect on science to an unprecedented degree, STSers will be able to reach new valuable insights into the actual workings of science. This might be useful also for the scientists themselves. After all, they are regularly asked by U.S. governmental bodies to specify criteria for scientific conduct—and misconduct, and may be interested in collaborating with sociologists to unravel existing standards and norms. It is from these kinds of mutually beneficial collaborations that some good will have eventually come out of the Science Wars.[26]

Still, there are larger issues at stake than the nature of scientific knowledge, or scientific standards and norms. Obviously the ability of scientists and sociologists of science to work together is crucial when it comes to solving real world problems which usually involve science and technology. Most issues today require *both* scientific and social scientific analysis, and here the social scientists, with their different expertise and orientation toward larger social mechanisms, have important contributions to make. This makes it even more important to establish an atmosphere of mutual understanding and respect beyond the Science Wars.

NOTES

1. The acronym STS stands for both Science and Technology Studies (an expansion of Science Studies or Social Studies of Science) and Science, Technology, and Society, certain interdisciplinary programs that were developed in the 1970s, some of which still are in operation. The latter type of STS typically was of interest also to practicing scientists and engineers, while the more esoteric academic pursuits of current science and technology studies have had an alienating effect on many earlier STS enthusiasts (see Bauer, chapter 2, and Fuller, chapter 9, for closer discussions of the history and changing nature of STS).

2. For instance, one of my political science colleagues coming back from a trip to Washington at the height of the Science Wars reported he had listened to a late night Public Television discussion there, where the Science Wars was presented just in this light.

3. Although the proscience activists in the Science Wars should not be assumed to be speaking for the majority of scientists, when it came to their opposition to social constructivism and relativism, however, they may, indeed, have represented a general feeling among scientists. The reasons will be examined later.

4. The case of the Unabomber cuts both ways, though, because of an early suggestion that the Society for the Social Studies of Science (4S) and the History of Science Society (HSS) may contain among its members professors who may have known and even inspired the Unabomber as a student. The leaders for both societies were interviewed by FBI at the joint 4S and HSS meeting in New Orleans, October 1994.

5. One of the reviewers of this book dismissed the view that the closedown of the supercollider "galvanized [the scientists] to defend their turf":

> This is too simplistic. Many scientists were opposed to the supercollider, for many reasons. One reason was that it was enormously expensive and would divert scarce resources from other scientific fields that were arguably more important (particle physics is considered arcane by many, including some physicists). Another reason was that it might

undermine international research efforts, such as CERN in Switzerland; it would be better to support CERN, or if a new supercollider was really needed, to build it in cooperation with other countries (e.g., Japan). There was rejoicing in scientific circles when the project was killed.

6. He also excluded the "freely suspended intelligentsia," that is, the intellectuals (Mannheim, 1936).

7. Among the programmatic references here are David Bloor's *Knowledge and Social Imagery* (1976, 1991), Andrew Pickering's *Constructing Quarks* (1984), Harry Collins's "The Empirical Program of Relativism" (1982) and *Changing Order* (1985), Bruno Latour and Steve Woolgar's *Laboratory Life* (1979, 1989), and Latour's *Science in Action* (1987).

8. The leading early laboratory studies were Bruno Latour and Steve Woolgar's *Laboratory Life* (1979, 1989) and Karin Knorr-Cetina's *The Manufacture of Knowledge* (1980), demonstrating at the microlevel how scientific facts were "socially negotiated"; see also Knorr-Cetina and Mulkay *Science Observed* (1983). At the Social Studies of Science Conferences, debates about social construction typically took place among others between philosopher Ronald Giere as a representative for "realism" and Pickering and Knorr as advocates of constructivism. Giere later contributed his book *Explaining Science* (1988) as his answer to this debate.

9. The Strong Programme was a self-conscious sociological attack on philosophic rationalist explanations or reconstructions of the history of science. The tenets of the Strong Programme were spelled out particularly in Bloor's famous *Knowledge and Social Imagery* (1976), which attempted to demonstrate that even mathematics and logics were social constructions. Early studies from Edinburgh typically emphasized the role of "social interests" in settling scientific controversies. Examples are the case studies in Barry Barnes's and Steven Shapin's (ed.) *Natural Order* (1979) and Donald McKenzie's *Statistics in Britain* (1981). For a discussion and critique of the Strong Programme, see particularly Laudan (1981), Woolgar (1981), and Schmaus, Segerstråle, and Jesseph (1992), "The Hard Program in the Sociology of Scientific Knowledge: A Manifesto." An early critique of the Strong Programme from the perspective of critical sociology was Daryl Chubin and Sal Restivo's "Weak Program" (Chubin and Restivo, 1983).

10. See e.g., Appleby, Hunt and Jacobs (1994) chapter 6, Gross and Levitt (1994), chapter 4, and Fuller, this volume, about the spread of French postmodernism in American academia.

11. This should not be construed so that the motivations of these two groups were the same. See Fuller, this volume, for a comment on the development in the humanities, and its separate life from that of science studies. It would also be incorrect to subsume the Science Wars under the Culture Wars, although there is clear overlap when it comes to what the philosopher John Searle in 1990 called "the cultural left" in an article discussing the Culture Wars on U. S. university campuses in the 1980s. For Searle, the cultural left consists of the new academics supporting such things as cultural studies, multiculturalism, "political correctness," and the like. Note that Searle in his seminal 1990 article had nothing to say

about social studies of science or the sociology of scientific knowledge as a political activity. It was, rather, Gross and Levitt (1994) who subsumed also social constructivists under the term "academic" or "cultural" left, thus making a connection not intended by Searle.

12. For more discussion, see Segerstråle (1993). An example of a study that takes scientists' convictions seriously is Segerstråle (2000).

13. At an earlier exhibition of the Enola Gay, veterans from World War II had come forth, outraged at the way in which the atom bomb was presented, not as the weapon that ended the war, but rather as responsible for the horrors of Hiroshima and Nagasaki.

14. For a discussion and criticism of "standpoint epistemologies," see Fuchs's chapter, this volume.

15. It should be noted, too, that over its almost three decades of existence, the sociology of scientific knowledge has moved away from its earlier emphasis on "interest" analysis, while sustaining its position of the social construction of all knowledge, including science (see, e.g., Shapin, 1995). Recently, there have been internal disagreements between various branches of SSK, as documented in Pickering (1992); it is, for instance, becoming clear that Latour is not a constructivist at all. Moreover, there have even been some "defections" from a strict social constructivist stance toward a recognition of some resistance from reality (Pickering, 1991, 1995). However, these developments, interesting as they may be, are of less consequence for the Science Wars, since the type of social constructivism that Gross and Levitt attack is the original core claim that scientific knowledge, just like all knowledge, is socially constructed.

16. We return to this question in chapter 4. Here we may note that there have indeed been attempts to educate the public exactly to make them make more responsible political decisions about science, such as in 1974 in Cambridge, Massachusetts, during the recombinant DNA controversy, and later governmental initiatives to educate citizens about nuclear power and other major policy issues in countries such as Sweden and Denmark.

17. Bauer (chapter 2, this volume) suggests that this may not have been a real problem, since if science was pitted against science studies, there was no doubt who would win.

18. This book was later translated into English under the title *Intellectual Impostures* (London: Profile 1998) and *Fashionable Nonsense* (New York: Picador 1998). See Dawkins (1998) for a positive review, Callon (1999) for a negative one.

19. Henry Bauer (personal communication) has informed me that he in vain tried to make Gross understand the difference between STS and various radical cultural movements. For Gross, the division line that matters is the one that goes between "constructivists" (both humanists and social scientists) and "ordinary" historians, sociologists, and philosophers of science. In his article, he explicitly excluded the latter from his criticism.

20. Surprisingly, STS seemed at the same time to be accused *both* of constructivism/postmodernism *and* of wanting to police science. However, a rampant constructivist or relativist (as imagined by Gross) would hardly have any Archimedean point from which to judge whether science lived up to certain stan-

dards, or would even have cared about such standards. Perhaps Gross implied that there after all existed different factions within STS, and that not everyone was constructivist/postmodern? If so, he would have been right. Or was this just a polemical move, aimed at giving scientists a number of different grounds for being angry with STS?

21. At that time Gross was also closer to the action; his 1997 article sported a Harvard e-mail address.

22. For instance, at the 1996 joint meeting of the Society for Social Studies of Science and The European Association for the Study of Science and Technology (EASST) in Bielefeld, Germany, petitions were circulated which suggested that some of the accusations of the critics of science studies were actionable (McIlwain, 1997).

23. Fuller (chapter 9) suggests that it is arguable whether one can see the Science Wars as a conflict between the Two Cultures. See, however, chapter 5 for an interpretation which restores the notion of a cultural conflict.

24. For an exception, see Restivo and Bauschpies (1996). This type of discussion, however, does go on outside the science studies community, in conjunction with such things as the Human Genome Project and cloning research. Also, as the reviewer in note 5 also pointed out, scientists themselves have taken these kinds of issues very seriously. He mentioned

> the revulsion of much of the scientific community to the atomic bomb, and the postwar division of the scientific community into two cultures, those who worked for the military directly or indirectly, and those who refused any participation in military work. Such questions as whether scientists should refuse to do certain research because of the uses to which the results might be put, or whether scientists should (or even were qualified to) become involved in policy-making to control those uses, emerged at that time and have not lost their urgency for scientists. This is one of the reasons that the attacks of the "antiscience warriors" meet with incomprehension on the part of scientists: where have they been?

25. In 1997 Harry Collins organized a conference at the University of Southampton (where he had moved from Bath) with an exchange between leading sociologists of scientific knowledge and leading scientists, particularly the physicists from the recent polemis in *Physics Today* and *Nature*. Collins referred to this as a "peace conference." This, however, seems like a misnomer as far as science studies were concerned. Obviously the Science Wars, was *not* between these (eminently reasonable) scientists and constructivist/relativist sociologists, but between certain *proscience* activists and the latter. In other words, in order to claim a "true" peace settlement in the Science Wars, Collins would have to find common ground with people like Gross and Levitt and Weinberg—or perhaps Wolpert?

26. As a defensive move—but also a move that reflects their self-perception—several representatives for the sociology of scientific knowledge in the mid-1990s declared that they themselves were *scientists* (Barnes, Bloor, & Henry, 1996; Collins, 1995; Pinch, 1995). The problem, however, was that natural scientists disagreed with the results of these "scientific" sociologists' research, particularly

their case studies (e.g., Labinger, 1995; Mermin, 1996a, b, c; Gottfried & Wilson, 1997). In principle at least, one outcome to this interesting situation might have been a serious "meta-discourse" about the nature of science—with a concomitant discussion about the exact nature of a desirable *sociology* of science.

In practice, however, this did not happen (yet?). As can be seen from assorted responses to Noretta Koertge's 1998 edited book *A House Built on Sand: Exposing Postmodern Myths about Science*, attacked SSK-sociologists have been busy playing down the significance of case studies, declaring their critics mistaken, or (sometimes justifiedly) complaining about misrepresentations by the "science police" (see e.g., Collins, 1999; MacKenzie, 1999a, b, responding to Sullivan, 1998, 1999; Pinch, 1999a, b, responding to McKinney, 1998, 1999; and Shapin and Schaffer, 1999a, b, responding to Pinnick, 1998, 1999). Nothing unusual is happening, from a sociology of science point of view. With abated breath, SSK-watchers wait for the next move of the Industry.

REFERENCES

Barnes, B., Bloor, D., and Henry, J. 1996. *Scientific Knowledge: A Sociological Analysis*. Chicago: University of Chicago Press.

Barnes, B. and Shapin, S. 1979. *Natural Order: Historical Studies of Scientific Culture* . Beverly Hills, CA: Sage

Beller, M. 1998. "The Sokal Hoax: At Whom Are We Laughing?" *Physics Today* (September): 29–34.

Benowitz, S. 1996. "Scientific Enterprise at Critical Juncture, Say Panelists, Researchers." *The Scientist*, October 14:3–5.

Bloor, D. 1976. *Knowledge and Social Imagery*. London: Routledge (2nd ed., 1991, Chicago: University of Chicago Press).

———. 1997. "Remember the Strong Program?" *Science, Technology and Human Values* 22 (3): 373–385.

Byerly, R., and Pielke, R. 1995. "The Changing Ecology of United States' Science." *Science* 269 (15 September):1531–1532.

Callon, M. 1999. "Whose Imposture?" Physicists at War with the Third Person. *Social Studies of Science* 29: 261–86.

Chubin, D., and Restivo, S. 1983. "The 'Mooting' of Science Studies: Research Programmes and Science Policy." Pp. 53–83 in Knorr-Cetina, K., and Mulkay, M., *Science Observed*. Beverly Hills, CA: Sage.

Cole, S. 1992. *Making Science*. Cambridge, MA: University of Chicago Press.

———. 1996. "Voodoo Sociology: Recent Developments in the Sociology of Science." Pp. 274–287 in P. Gross, Levitt, N., and Lewis, M. (eds.), *The Flight from Science and Reason*. Baltimore, MD: Johns Hopkins University Press.

Collins, H. 1982. "Knowlege, Norms and Rules in the Sociology of Knowledge." *Social Studies of Science* 12:299–309.

———. 1985. *Changing Order: Replication and Induction in Scientific Practice*. Beverly Hills, CA: Sage.

———. 1999. "The Science Police." Review of Noretta Koertge (ed.), *A House Built on Sand: Exposing Postmodernist Myths about Science. Social Studies of Science*

29: 287–94.
———, and Pinch, T. 1993. *The Golem: What Everybody Should Know about Science.* Cambridge: Cambridge University Press.
———, and ——— 1996. Response to Mermin (letter). *Physics Today* (July): 11, 13.
———, and ——— 1997. Reply to letters from scientists. *Physics Today* (January):15, 92,94.
———, and Yearley, S. 1992a. "Epistemological Chicken." Pp. 301–326 in Pickering, A. (ed.), *Science as Practice and Culture.* Chicago: University of Chicago Press.
———, and ——— 1992b. "Journey into Space." Pp. 369–389 in Pickering, A. (ed.), *Science as Practice and Culture.* Chicago: University of Chicago Press.
Collins, R., and Restivo, S. 1983. "Development, Diversity and Conflict in the Sociology of Science." *The Sociological Quarterly* 24: 185–200.
Dawkins, R. 1994. "The Moon is Not a Calabash." *Times Higher Literary Supplement,* September 30:17.
Dawkins, R. 1998. "Postmodernism Disrobed." Review of Sokal and Bricmont, *Intellectual Impostures. Nature* 394: 141–3.
Edge, D. 1996a. "Stop Knocking Social Sciences." Letter. *Nature* 384:106.
———. 1996b. Editorial. *Social Studies of Science* 26:5–6.
———. 1996c. Editorial. *Social Studies of Science* 26:723–732.
———. 1996d. Letter. *Science* 274 (8 November):904
———. 1997a. "Beam and Mote." Letter to *Nature* (unpublished).
———. 1997b. "For the Record." *Social Studies of Science* 27: 357–359.
———. 1997c. "For the Record." *Social Studies of Science* 27: 543–544.
Fish, S. 1996. "Profesor Sokal's Bad Joke." *The New York Times,* May 21, A 11.
Flam, F. 1994. "Privately Funded Exhibit Raises Scientists' Ire." *Science* 265:729.
Forman, P. 1995. "Truth and Objectivity." *Science* 269:565–567 and 707–710.
Forman, P. 1997a. "Assailing the Seasons," Review of Gross, P. Levitt, N., and Lewis, M. (eds.) *The Flight from Science and Reason.* Science 276 (2 May): 750–752.
———. 1997b. Response. *Science* 276 (27 June):1955.
Fraser, S. (ed.). 1995. *The Bell Curve Wars.* New York: Basic Books.
Fuchs, S. 1992. *The Professional Quest for Truth: A Social Theory of Scientific Knowledge.* Albany: SUNY Press.
Fuller, S. 1994. "Can Science Studies Be Spoken in a Civil Tongue?" *Social Studies of Science* 24 (1) February:143–168.
———. 1995. "Two Cultures II: Science Studies Goes Public." *EASST Newsletter,* Spring.
Geison, G. 1995. *The Private Science of Louis Pasteur.* Princeton, NJ: Princeton University Press.
Gibbons, M., Limoges, C., Nowotny, H., Schwartzman, S., Scott, P., and Trow, M. 1994. *The New Production of Knowledge: The Dynamics of Science and Research in Contemporary Science.* London: Sage.
Giere, R. 1992. "The Cognitive Construction of Scientific Knowledge" (Response to Pickering). *Social Studies of Science* 22:95–107.
———. 1988. *Explaining Science.* Chicago: University of Chicago Press.

Gieryn, T. 1996. "Policing STS: A Boundary-Work Souvenir from the Smithsonian Exhibition on 'Science in American Life.' " *Science, Technology and Human Values* 21 (1) Winter:100–115.

Gottfried, K. 1997. "Was Sokal's Hoax Justified?" *Physics Today* (January):61–62.

———, and Wilson, K. 1997. "Science as a Cultural Construct." *Nature* 386 (10 April):545–547

Gross, P. 1996 "Reply to Tom Gieryn." *Science, Technology and Human Values* 21 (1) (Winter):116–120.

———. 1997. "Opinion: The So-Called Science Wars and Sociological Gravitas." *The Scientist*, April 28:8.

———, and Levitt, N. 1994. *Higher Superstition: The Academic Left and Its Quarrels with Science*. Baltimore, MD: Johns Hopkins.

———, and ———. 1995. "Knocking Science for Fun and Profit." *The Skeptical Inquirer* 19 (2) March/April:38–42.

———, Levitt, N., and Lewis, M. (eds.). 1995. "The Flight from Science and Reason." *Annals of the New York Academy of Sciences* (775), 24 June.

———, ———, and ———. (eds.). 1996. *The Flight from Science and Reason*. Baltimore, MD: Johns Hopkins University Press.

Gruner, S., Langer, J., Nelson, P. and Vogel, V. 1995. "What Future Will We Choose for Physics?" *Physics Today* (December):25–30.

Gunsalus, C. K. 1997. "Opinion: Rethinking Unscientific Attitudes about Scientific Misconduct." *Chronicle of Higher Education*, March 28.

Guston, D. 1995. "The Flight from Reasonableness." (Report from the "The Flight from Science and Reason" conference.) *Technoscience* 8 (3), Fall:11–13.

Hagendijk, R. 1990. "Structuration Theory, Constructivism, and Scientific Change." Pp. 43–66 in Cozzens, S. E., and Gieryn, T. F. (eds.), *Theories of Science in Society*. Bloomington and Indianapolis: Indiana University Press.

Harding, S. 1991. *Whose Science? Whose Knowledge*? Ithaca, NY: Cornell University Press.

Heller, S. 1994. "At Conference, Conservative Scholars Lash Out at Attempts to 'Delegitimize Science.' " *The Chronicle of Higher Education*, November 23: A18, A20.

Herschbach, D. 1997. Letter. *Science* 276 (27 June):1954.

Herrnstein, R., and Murray, C. 1994. *The Bell Curve: Intelligence and Class Structure in American Life*. New York: Free Press.

Hollinger, D. 1995. "Science as a Weapon in Kulturkampfe in the United States during and after World War II." *Isis* 86 (September):440–454.

Holquist, M., and Shulman, R. 1996. Letter. *The New York Review of Books*, October 3: 54.

Holton, G. 1993. *Science and Anti-Science*. Cambridge, MA: Harvard University Press.

Horgan, J. 1996. *The End of Science*. New York: Addison-Wesley.

Irwin, A. and Wynne, B. 1996. *Misunderstanding Science? The Public Reconstruction of Science and Technology Studies*. Cambridge: Cambridge University Press.

Jasanoff, S. 1995. "Cooperation for What?: A View from the Sociological/Cultural Study of Science Policy." *Social Studies of Science* 25 (2):314–317.

———. 1996. "Beyond Epistemology." *Social Studies of Science* 26:393–418.

———, Markle, G., Petersen, J. and Pinch, T. 1995. *Handbook of Science and Technology Studies*. Thousand Oaks, CA: Sage.
Koertge, N. 1998. *A House Built on Sand*. Oxford: Oxford University Press.
Kuhn, T. 1970. *The Structure of Scientific Revolutions*. Chicago: University of Chicago Press.
Labinger, Jay. 1995. "Science as Culture: A View from the Petri Dish." *Social Studies of Science* 25:285–306.
LaFollette, M. 1995. "Wielding History Like a Hammer." *Science Communication* 16:235–241.
Latour, B. 1987. *Science in Action*. Cambridge, MA: Harvard University Press.
———. 1992. "One More Turn after the Social Turn." Pp. 272–294 in McMullin, E. (ed.), *The Social Dimensions of Science*. Notre Dame: University of Notre Dame Press.
———. 1996. *Aramis or the Love of Technology*. Cambridge, MA: Harvard University Press.
Laudan, L. 1981. "The Pseudo-Science of Science." *Philosophy of the Social Sciences* 11: 173–198; reprinted in Brown, J. R. (ed.) *Scientific Rationality: The Sociological Turn* (Dordrecht: D. Reidel. 1984).
Levine, G. 1996. Letter. *The New York Review of Books*, October 3:54.
Levitt, N. 1997. Letter. *Science* 276 (27 June):1953.
———, and Gross, P. 1994. "The Perils of Democratizing Science." *The Chronicle of Higher Education*, 5 October:B1–B2.
Longino, H. 1990. *Science as Social Knowledge*. Princeton: Princeton University Press.
Lynch, W. 1994. "Ideology and the Sociology of Scientific Knowledge." *Social Studies of Science* 24: 197–227.
MacKenzie, D. 1999a. "Comment" The Science Wars and the Past's Quiet Voices." *Social Studies of Science* 29: 199–213.
MacKenzie, D. 1999b. "The Zero-Sum Assumption: Reply to Sullivan." *Social Studies of Science* 29: 223–34.
Magyar, G. 1996. "Mote and Beam." Letter. *Nature* 384 (December 12):506.
Martin, B. 1996. "Social Construction of an 'Attack' on Science." (Review of Gross and Levitt). *Social Studies of Science* 26:161–173.
McKinney, W. J. 1998. "Where Experiments Fail: Is 'Cold Fusion' Science as Normal?" Pp. 133–50 in Koertge, N. (ed.), *A House Built on Sand: Exposing Postmodernist Myths about Science*. New York and Oxford: Oxford University Press.
McKinney, W. J. 1999. "Partial Houses Built on Common Ground." *Social Studies of Science* 29: 240–246.
McMillen, L. 1997. "The Science Wars Flare at the Institute for Advanced Study." *Chronicle of Higher Education*, May 16:A13.
Mermin, D. 1996a. "What's Wrong with This Sustaining Myth?" (Review of Collins and Pinch's *The Golem*.) *Physics Today* (March):11, 13.
———. 1996b. "The Golemization of Relativity." *Physics Today* (April):11, 13.
———. 1996c. Reply to Collins and Pinch. *Physics Today* (July):13, 15.
———. 1997. Reply to Collins and Pinch. *Physics Today* (January):94.
Merton, R. K. 1942/1973. "The Normative Structure of Science." Pp. 267–278 in

Merton, R. K., *The Sociology of Science*. Chicago: The University of Chicago Press. (Republication of original, entitled "Science and Technology in a Democratic Social Order," *Journal of Legal and Political Sociology* 1 (1942): 115–26.

Mulkay, M. 1997. Review of Irwin and Wynne: Misunderstanding Science? *Science Technology and Human Values* 22 (2):254–258.

Nelkin, D. 1996a. "Responses to a Marriage Failed." *Social Text* 1/2 (Spring/Summer):93–100.

———. 1996b. "What Are the Science Wars Really About?" Opinion. *The Chronicle of Higher Education*, July 26:A 52.

Perutz, M. 1996. Review of Gerald Geison: *The Private Life of Louis Pasteur. New York Review of Books*, December 21.

———. 1997. Reply to Summers. *New York Review of Books*, February 6:41–42.

Pickering, A. 1984. *Constructing Quarks*. Chicago: University of Chicago Press.

———. 1991a. "Philosophy Naturalized a Bit." *Social Studies of Science* 21:575–584.

———. 1991. "Objectivity and the Mangle of Practice." *Annals of Scholarship* 8:409–425.

———. (ed.) 1992. *Science as Practice and Culture*. Chicago: University of Chicago Press.

———. 1995. *The Mangle of Practice: Time, Agency, and Science*. Chicago: University of Chicago Press.

Pinch, T. 1999a. "Half a House: A Response to McKinney." *Social Studies of Science* 29: 235–40.

Pinch, T. 1999b. "Final Response to McKinney." *Social Studies of Science* 29: 246–47.

Pinnick, C. 1998. "What's Wrong with the Strong Programme's Case Study of the 'Hobbes-Boyle' Dispute?" Pp. 227–39 in Koertge, N. (ed.), *A House Built on Sand: Exposing Postmodernist Myths about Science*. New York and Oxford: Oxford University Press.

Pinnick, C. 1999. "Caught in a Sandy Shoal of the Shallow: Reply to Shapin and Schaffer." *Social Studies of Science* 29: 253–57.

Restivo, S. 1995. "The Theory Landscape in Science Studies." Pp. 95–110 in Jasanoff, S., Markle, G., Petersen, J., and Pinch, T. (eds.), *Handbook of Science and Technology Studies*. Thousand Oaks, CA: Sage.

Restivo, S., and Bauschpies, W. 1996. "Science, Social Theory, and Science Criticism." *Communication and Cognition*. 29 (2):249–272

Rose, H. 1996 "My Enemy's Enemy Is—Only Perhaps—My Friend." *Social Text* 1/2 (Spring/Summer):61–80.

Ross, A. 1996. *The Science Wars*. Durham, NC: Duke University Press.

Schmaus, W., Segerstråle, U., and Jesseph, D. 1992. "The Hard Program in the Sociology of Science: A Manifesto." *Social Epistemology* 6 (3):243–265. (With peer commentaries.)

Schweber, S. "Reflections on the Sokal Affair: What is at Stake?" *Physics Today*, March 1997: 73–74.

Searle, J. 1990. "The Storm Over the University." *New York Review of Books*, December 6.

Segerstråle, U. 1993. "Bringing the Scientist Back In-The Need for an Alternative

Sociology of Scientific Knowledge." Pp. 57–82 in Brante, T., Fuller, S. and Lynch, W. (eds), *Controversial Science*. Albany, NY: SUNY Press.

———. 1994. "Science by Worst Cases." (Book review of H. Collins and T. Pinch, *The Golem: What Everyone Should Know About Science*.) *Science* 263 (11 February):837–838.

———. 1995. "Good to the Last Drop? Millikan Stories as 'Canned' Pedagogy." *Science and Engineering Ethics* 1 (3):197–214.

Segerstråle, U. 2000. *Defenders of the Truth: The Battle for Science in the Sociobiology Debate and Beyond*. Oxford and New York: Oxford University Press.

Shapin, Steven. 1994. *A Social History of Truth*. Chicago, IL: University of Chicago Press.

———. 1995. "Here and Everywhere: Sociology of Scientific Knowledge." *Annual Review of Sociology* 21:289–321.

———, and Schaffer, S. 1985. *Leviathan and the Airpump*. Princeton, NJ: Princeton University Press.

———, and Schaffer, S. 1999a. "Response to Pinnick." *Social Studies of Science* 29: 261–86.

———, and Schaffer, S. 1999b. "On Bad History." Response to Pinnick. *Social Studies of Science* 29: 257–59.

Slaughter, S., and Leslie, L. 1997. *Academic Capitalism: Politics, Policies, and the Entrepeneurial University*. Baltimore, MD: Johns Hopkins University Press.

Snow, C. P. 1959. *The Two Cultures*. Cambridge: Cambridge University Press.

Sokal, A. 1996a. "Trangressing the Boundaries: Toward a Transformative Hermeneutics of Quantum Gravity." *Social Text* 46–47: 217–252.

———. 1996b. Sokal's response. *Lingua Franca* 6 (4):62–64.

Sokal, A., and Bricmont, J. 1997. *Impostures Intellectuelles*. Paris: Odile Jacob.

Sokal, A., and Bricmont, J. 1998. *Intellectual Impostures*. London: Profile.

Sokal, A., and Bricmont, J. 1998. *Fashionable Nonsense*. New York: Picador.

Stephan, P. 1996. "The Economics of Science." *Journal of Economic Literature* 34:1199–1235.

Sullivan, P. A. 1998. "An Engineer Discusses Two Case Studies: Hayles on Fluid Mechanics and MacKenzie on Statistics." Pp. 71–98 in Koertge, N. (ed.), *A House Built on Sand: Exposing Postmodernist Myths about Science*. New York and Oxford: Oxford University Press.

Sullivan, P. A. 1999. "Response to MacKenzie." *Social Studies of Science* 29: 215–23.

Summers, W. 1997. "Pasteur's Private Science." Letter. *New York Review of Books*, February 6:41.

Trefil, J. 1997. Letter. *Science* 276 (27 June):1953.

Weinberg, S. 1992. *Dreams of a Final Theory*. New York: Pantheon; London: Hutchinson.

———. 1994. "Response to Steve Fuller." *Social Studies of Science* 24 (4) November:748–751.

———. 1995. "Night Thoughts of a Quantum Physicist." *Bulletin of the American Academy for Arts and Sciences* 29 (3) December:51–64.

———. 1996a. "Sokal's Hoax." *The New York Review of Books*, August 8.

———. 1996b. "Response." *The New York Review of Books*, October 3: 55–56.

Wise, N. 1996. "The Enemy Without and the Enemy Within." *Isis* 87 (2) June:323–327.

———. 1996. Letter. *The New York Review of Books*, October 3:54.

Wolpert, L. 1992. *The Unnatural Science of Science: Why Science Does Not Make (Common) Sense*. London: Faber & Faber (Cambridge, MA: Harvard University Press, 1993).

———. 1994. "Response to Steve Fuller." *Social Studies of Science* 24 (4) November:745–747.

Ziman, J. 1994. *Prometheus Bound: Science in a Dynamic Steady State*. Cambridge: Cambridge University Press.

Zürcher, R. 1996. "Farewell to Reason: A Tale of Two Conferences." *Academic Questions* 9 (2): 52–60.

CHAPTER TWO

Antiscience in Current Science and Technology Studies

HENRY H. BAUER

EDITOR'S INTRODUCTION

Henry Bauer's chapter represents a cri de coeur *by someone who is both a scientist and a veteran of science studies. This chapter was written at the height of the Science Wars. As a science studies practitioner, Henry Bauer has himself in books and articles explored such things as the murky boundaries between science and pseudoscience, and the public understanding of science (in* Scientific Illiteracy and the Myth of the Scientific Method*). Therefore, when Bauer in this chapter criticizes his science studies colleagues, he does it for rather different reasons than Gross and Levitt (with whom he does not always agree). Bauer is irritated with the current practitioners of (STS): he feels they are undermining the* raison d'être *for their own field—his field, too—which he sees as the* social *study of science. He contrasts the recent type of research with the* original *idea of STS—Science, Technology and Society—which he found so fascinating that it lured him away from chemistry! The earlier type of STS critically examined the relationship between science and society without making extreme epistemological assumptions about science. Scientists, in fact, took the initiative to this kind of STS and were active in the field, and members of the 4S society; now they feel alienated.*

Intellectually, Bauer's dissatisfaction with the dominant paradigm in the new STS has to do with the epistemological aim of the new science studies, which is to show that there is no difference between science and other human activities. As a scientist, Bauer wants to find out how reliable knowledge has been attained nevertheless. And he certainly puts his foot down in regard to a minimal criterion of validity for science studies: The result should not seem absurd to thoughtful, philosophically minded scientists! *He concludes that sheer self-preservation of STS requires that it make peace with science on terms congenial to science and its publics.*

RECENT DEVELOPMENTS

Antagonism toward science has not been prominent until rather recently among those engaged in science and technology studies. Indeed, scientists themselves were early participants in the several streams that seem now to have joined into "science studies" or "STS." To find science currently attacked from within STS is puzzling to scientists: it seems perverse for people to despise the subject of their own scholarship; and it is also rather like biting the hand that feeds, or killing the goose that lays the golden eggs. It is also curious to find purportedly sophisticated students of science propounding major misconceptions about it.

On the whole, the scientific community has paid little attention to STS. Moreover, scientists are in the habit among themselves of deferring to experts and have presumed that philosophers, historians, and sociologists of science surely know what they are talking about. Scientists began to react openly against what they considered snide denigration and overt attacks in 1994, with the publication of Gross and Levitt's *Higher Superstition*. The backlash has since grown. Academe is increasingly pressed by budgetary restrictions and external urgings to concentrate more on teaching and less on research. The scientific community, in particular, is being charged to refurbish its ethics. So internal, external, and mutual critiquing and jockeying for preferment are likely to be troublesome for some time to come. As science sets out to defend itself, critics of science will find themselves increasingly opposed with increasing forcefulness.

It has always been necessary for the scholarly health of STS that its aim be to *understand* science. Now that may become a matter of self-preservation. The scientific community enjoys far higher societal status than does STS, social science, or the humanities. If the public image of STS becomes synonymous with that of relativism and antagonism toward science, the institutional progress that STS has made over the years may well be reversed.

IS THERE REALLY ANTAGONISM AGAINST SCIENCE IN STS?

The Fact of Antiscientific Sentiment

The fact of antiscientific sentiment has been acknowledged even by as sympathetic a science studies insider as John Ziman: in contemporary STS, "much discourse . . . seems aimed at proving that . . . [science] works . . . on behalf of certain sinister power groups (Ziman, 1994, p. 275).

For scientists looking at current STS from the outside, it is of course the loudest voices that are most noticed. Gross and Levitt go even as far as equating STS with the radical ideologues of animal-rights activism, feminism, Marxism, and sociological relativism like that of the "strong programme" (Gross & Levitt, 1994). The overgeneralization of Gross and Levitt may already have been accepted in some quarters. For instance, a statement such as "[S]cientists also should confront the sociologists and philosophers at their institutions who are attacking the foundations of science" may or may not refer to all sociologists and philosophers (Bard, 1996). There is little doubt, however, of such a generalization when, commenting on the recently revised history standards, a leader of the American Physical Society conceded that "they did a pretty good job . . . I'm really quite astonished that they did as good a job as they did considering the attitude of historians toward science"—not "some historians," just "historians" (Robert Park cited in Holden, 1996).

Actual events lend support to such perceptions among scientists. A few years ago, the American Chemical Society (ACS) donated $5.3 million for the Smithsonian Institution to mount an exhibit on science in American life intended inter alia to stimulate interest in careers in science among girls and children of minority groups—"the largest financial donation for a single exhibit in the history of the Smithsonian" (Ross, 1994). No sooner was the exhibit opened than prominent scientists began to express dismay: "Ring the bell of evil, and viewers will automatically blame a scientist" (Nemecek, 1995). Many scientists saw the exhibit as a very negative statement about science. According to Marcel LaFollette, herself an STS insider, "It is . . . remarkable then that the ACS itself became the target of so much internal hostility from museum staff. Several historians . . . made no secret of their disdain for 'Big Science' and of everything they believed it represented . . . They wound up creating a largely negative [exhibit]" (LaFollette, 1995). LaFollette reports seeing the majority of ACS scientists change from respect for the Smithsonian to disillusion after *years* of "contentious debate"; "an ACS advisor struggled to maintain his composure while a Smithsonian historian harangued him about the horrors of agricultural chemicals. The curator, determined to press a political argument, seemed smugly unconcerned about the effect of his rudeness." After more than a year of negotiating for changes, the ACS "told the museum that further negotiations would be fruitless."[1]

The scientific community will not acquire respect for the science studies community when it sees such intransigence and inability to communicate with leading organizations of science; or when an assistant professor of sociology at one of the major centers of STS finds space in the *Chronicle of Higher Education* to state that "Today's scientists are finding it difficult to deal with the fact that they no longer automatically inherit the near-godlike status that many American scientists were accorded in the years following World War II" (Kleinman, 1995). It is not easy for one who through actually doing science has learned to base opinion on evidence, to excuse such a sweeping overgeneralization, or the assertion that scientists' response to the recent attacks is "undoubtedly [*sic*] . . . an effort to protect the unique status that science and scientists have enjoyed"; or the suggestion that *not* "only certified scientists should have any role . . . in making decisions concerning science—including the research methods employed." Suspicion of science is also displayed when "such high-profile disasters as those at Love Canal and Three Mile Island" are associated with an imagined "right of scientists to work without public scrutiny."

And it is little wonder that leading proponents of constructivism are judged by the scientific community to be both ignorant and antagonistic when they claim, for example, that "experiments are rarely decisive in settling scientific controversies" (Collins & Pinch, 1993; for a critique, see Segerstråle, 1994). In actual fact, of course, experiment or observation—at any rate empirical evidence—is the only way in which scientific disagreements ever do get finally resolved. The shibboleth of "experimenter's regress" stems from studies of controversies made *before they have been finally settled,* for instance as to the detection of gravity waves. Permanently settled matters in science—say, the sequence and grouping of elements in the Periodic Table—continue to be agreed to be settled only insofar and for as long as they have adequate empirical justification.

Though the backlash against academic antiscience is associated with Gross and Levitt, they only brought to public attention what had been brewing for some time. In 1988, Lewis Wolpert commented about Woolgar's *Science: The Very Idea* that it "seems a pity that such intelligence has been used to attack science rather than provide badly needed insights into science and its relation to society," with an accompanying photo of Woolgar labeled "HOSTILE: The author, Woolgar" (Wolpert, 1988). An accompanying review of Latour's *The Pasteurization of France* described it as "Starting out as history of science, it gradually becomes science fiction . . . this book is great fun to read, once one decides that it has nothing to do with science (Brock, 1988).

Antiscientific Sentiment in Perspective

Societal antagonism toward science waxes and wanes. Though both tendencies are always present, historians have discerned cycles of dominant rationalism and dominant antirationalism (or Romanticism) over a millennium or more of Western civilization. Stephen Brush has explicitly traced cycles of Romanticism and realism since the Scientific Revolution (Brush, 1978). Julius Greenstone finds cycles of rationalism and mysticism in the history of Messianic notions in Judaism (Greenstone, 1972).[2]

The Scientific Revolution in seventeenth-century western Europe is often taken as harbinger of the eighteenth-century Enlightenment in which rationality in all matters became a guiding principle; used, abused, or misused in the service of social revolution, for example, in France. The intellectual "excess of abstraction that marked the end of the Age of Reason" stimulated the reaction of early-nineteenth-century Romanticism with its emphasis on "concreteness and the love of common facts" (Barzun, 1994). The middle and late nineteenth century then saw a resurgence of "realism" in response to an excess of Romanticism. At the end of the nineteenth century another predominantly negative response to realism came in neoromanticism; to be followed in turn by a period of *neorealism* (Brush, 1978)

Periods of rapid scientific advance or dominant rationalism seem often to arouse Romantic notions of an anti- or pseudo-scientific sort (Bauer, 1986–1987). Jonathan Swift illustrates the intellectual reaction to overweening scientism in the eighteenth century.[3] Nineteenth-century discoveries in electromagnetism were accompanied by a plethora of electric health-cures and quackery. The realism of natural selection and the tracing of human origins to the animal kingdom came simultaneously with Spiritualism and the enthusiastic investigation of psychic phenomena. The end of the nineteenth century and the beginning of the twentieth saw not only the genuine discoveries of radioactivity and X rays but also the illusory ones of N rays and mitogenetic radiation. World War II was followed not only by an avalanche of new science and new technology but also by the advent of flying saucers and renewed interest in cryptozoology and parapsychology. The present, most modern scientific age also harbors New Age beliefs that counter the cold, hard facts of established science with intuitive, mystical, pantheist concepts; and it also harbors intellectual reaction against scientism, exemplified, say, by Paul Feyerabend, in which the determination to resist scientism becomes reckless, overgeneralizing denigration of anything and everything to do with science.

Early in this postwar period, C. P. Snow pointed to the intellectual chasm that had opened between the scientific and the nonscientific intellectual realms. He was prescient in warning that "incomprehension of science . . . gives an unscientific flavour to the whole 'traditional' culture . . . that . . . is often, much more than we admit, on the point of turning antiscientific" (Snow, 1963:17).

That cycles of pro- and antiscience are endemic in society, that antagonism persists even when adulation is dominant, makes it neither necessary nor proper, however, for academic students of science to vent animus against it. Scientists and the wider public believe that academic study should be *dispassionate, disinterested, apolitical.* It does not sit well when activists claim academic freedom to push quite blatantly their political agenda, as the feminist and relativist science-criticizers commonly do.

Incomprehension of Science

The "strong programme" and its ilk are fundamentally and irretrievably wrongheaded, if the purpose is to understand science. So is its primary "principle," that scientific activity should be investigated "neutrally," without prior judgment about the correctness or incorrectness of the science being studied. As happens so often, something that sounds reasonable in the abstract becomes ridiculous when applied to any real case. So far as *science* is concerned—that is, substantive knowledge about Nature—it matters crucially whether data or theories fit or do not fit, are operationally right or wrong. A nice illustration of the absurdity that ensues when "right or wrong" is said not to matter in the assessment of purportedly scientific controversy can be found in the Velikovsky Affair (Bauer, 1984a). The replay of that 1950s controversy in the early 1960s and again in the early 1970s was fueled by social scientists who dwelled on how badly the individual, Velikovsky, had been treated, asserting that fairness and openness to new ideas was owed in science to anyone, irrespective whether the ideas turned out ultimately to be right or wrong. Under this type of argument scientists would be endlessly doomed to examine the propositions of flat-earthers, 6,000-year creationists, and the like just so long as the proponents are articulate and intelligent albeit ignorant (Bauer, 1984b).

A common way in which some science studies scholars foster incomprehension of science is to focus on one aspect of it to the exclusion of all else: describing science as "business in disguise," say, or "deconstructing" a celebration of X ray crystallography as a "succession rite."[4] That is no more valid than the old-fashioned scientistic showing how scientists seek disinterestedly to serve humankind. Both capture only part of the picture; and the motives of those who look only at the unflattering side of science are at least as suspect as are those of the ones who only flatter.

That sort of one-dimensional, externalist commentary requires no technical knowledge of science itself. This raises the question, How much or how little does an interpreter need to know to be able to speak validly and significantly? Maybe STS should follow the example of traditional history of science and define itself as a field of graduate study open only to people who already have degrees in science. But much more immersion in science than that is needed before one has an authentic feel for it. Undergraduate courses in science typically focus, necessarily, on the reliable stock of textbook science rather than on the uncertainties encountered in *frontier* science (Bauer, 1986, 1992). Practicing scientists encounter the realities of knowledge-seeking only when they do research themselves. Only then do they get the personal experience that may lead to a balanced understanding of the shadings of reliability and unreliability that characterize different bits of "scientific knowledge." Ideally, then, those who practice STS would themselves have had significant experience of doing scientific research. Failing that, however, they should at least pay considerable heed to those few outstanding scientists who, in addition to their scientific careers, have also become competent in STS. Perhaps the most telling criticism that could be made of the antiscientistic extremists in STS is that they ignore the veritable, authoritative insights about science that permeate the work of persons like Michael Polanyi and John Ziman.

Might one not adopt as a criterion of validity in "science studies" that the result not seem absurd to thoughtful, philosophically minded practicing scientists? A not inconsiderable quantity of spoken and published STS would not pass muster under that criterion. Many examples of such wrongheaded absurdities are cited by Gross and Levitt in *Higher Superstition*. Among those that have astonished me personally: Steven Shapin saying that scientists do not value their technicians, and Andrew Pickering seeing no compelling reason to exclude the possibility of the future discovery of a chemical element between hydrogen and helium.[5] Anthropologists should have learned from the Margaret Mead Affair that the natives ought to be explicitly consulted before conclusions arrived at by anthropologists can be accepted (Freeman, 1983). Anthropological approaches in STS suffer just the same hazard, that outsiders can be wildly, absurdly wrong in their conclusions.

Hypocrisy: Preaching versus Practice

When push comes to shove, though, many radical science studies scholars seem to lack the courage of their epistemological convictions, or rather the conviction of their assertions. The slipperiness is illustrated by a recent sidebar vignette in *Science*, where we find the following quote of a cultural constructivist:

> In the numerous conversations I have had with scientists about social constructivism, gravity is invariably brought forward as the great counterexample showing that science is not culturally constructed. A rock falls to earth regardless of the dominant language or ruling class. *Yet even the pervasiveness of this example indicates that it is culturally encoded . . . marked by a certain cultural position because it presupposes that mathematics and physics are the core sciences rather than, say, biology and ecology.*[6] (Emphasis added)

But so far as scientists are concerned, the argument over "culturally constructed," is *purely* whether or not the behavior of bodies is accurately described under the theory of gravity. That's all. No more than that. But also certainly not less. That a scientist's *choices of example* in arguments with nonscientists are culturally influenced is an entirely other matter, irrelevant to the central issue. Furthermore, what the "core sciences" might be seen to be is again an other matter—not incidentally, one that is of little or no interest to working scientists, most of whom are preoccupied with their technical tasks and could not care less what somebody may think is the hierarchy of the sciences.

The relativists must indulge in such tricky footwork because otherwise they admit that they have nothing useful to say. If science's pronouncements are mere expressions of ideology or self-interest, then those of the relativists are too. Gross and Levitt have described quite authentically the desire of constructivists and relativists to evade the consequences of their stated beliefs: "[C]onstructivism . . . is . . . relentlessly mechanistic and reductionist . . . all are puppets of the temper of an age. . . . Only the cultural constructivists themselves (of course) are licensed to escape the intellectual tyranny of this invisible hand. . . . Typically, in the face of all-out challenges from scientists and philosophers . . . they edge away from the strong version of the constructivist claim . . . [but with] a different audience, one primed to hear science contextualized, relativized, and revealed as the deformed offspring of capitalist hegemony, the constructivist claws come out once more" (Gross & Levitt, 1994, p. 56–57).

In two long essay-reviews, Paul Forman has given chapter and verse of the retreat from the strong version of their views by prominent constructivists, relativists and feminism-theorists *who fail to acknowledge the implications of that retreat* (Forman, 1995a,b). Their dilemma is plain enough, of course. To say that science is embedded in a wider society that exerts a certain influence on how it functions is sleep-inducingly uncontroversial; the most naive and scientistic scientist would agree, well aware of where research funds come from, or the persistent efforts of creationists to get into the biology classrooms and textbooks, and so on. Not much notice would ever have been taken had the Edinburghers and their fellow-travelers made this their "programme" and had studied in specific instances how societal influence played out in the elucidation of nevertheless rather reliable scientific knowledge. Historians of science, after all, have long known that external as well as internal factors need to be taken into account for any genuine, complete understanding of how science has wrought. Thomas Kuhn's scenario of "scientific revolutions" complete with "incommensurability" and the like quickly found meaningful resonance with scientists. It is only the most extreme, know-nothing epistemological relativism that the scientific community will not stomach.

THE PURPOSE OF STS

How Did STS Come into Being?

STS derives from several initiatives that have by no means come fully together.

Concern among scientists that science should purposely contribute to the common welfare was exemplified in 1930s Britain by J. D. Bernal, whose personal political convictions lent a Marxist flavor to the movement in which he was so prominent.[7] In the United States, a similar movement but without the overt Marxism led to the founding (and flourishing) after World War II of the *Bulletin of the Atomic Scientists*. The Pugwash conferences were another manifestation of scientists' concern to engage in policy discussion. Some of the presently existing STS programs with a strong orientation toward policy studies are direct descendants of these precursors. Within the humanities one can discern at least two distinct sources of STS. The intellectual impetus is often and plausibly traced to Kuhn's argument for a realistic, historically based philosophical understanding of science (Kuhn, 1970). The subsequent and continuing appearance (though not always the durability!) of multi- or interdisciplinary centers or departments—History and Philosophy of Science, or Philosophy and Sociology of Science, and the like—attests the recognition that an adequate understanding of science in its societal context requires that philosophers of science join with historians, sociologists, and others in pursuit of explanations for the success of science and of proper criteria for such success. Full-blooded attempts at inter- or multidisciplinary scholarship and teaching led to the founding of Units or Departments or Centers of what has come to be called "Science Studies" in Britain or "STS" (Science, Technology, and Society) in the United States. Professional associations too reflect these attempts at synthesis, as for example the Society for Social Studies of Science (founded in the 1970s) or the Society for History, Philosophy, and Social Studies of Biology (established in the 1980s). An attempt to survey the intellectual ramifications of policy as well as of academic studies was made by Spiegel-Rösing and de Solla Price in 1977 (Spiegel-Rösing & de Solla Price, 1977).

Contemporaneous with that intellectual initiative was the student-led activism of the 1960s, part of which turned to criticism of science as a culprit (or even *the* culprit) in certain unsatisfactory aspects of technological society. Innumerable undergraduate courses, sometimes more or less organized into programs, dealt with issues of "science and society": environmental pollution, problems of nuclear energy, and so on.[8] Some current STS programs derive from these ancestors.

STS exists, then, for several reasons that are not necessarily compatible. Most philosophers and historians and some sociologists, scientists, and others want it to be a predominantly intellectual venture aimed at the most complete possible understanding of the ramifications and implications of science. Some scientists, political scientists, and others are primarily concerned that public policy take adequate account of the insights and capabilities that science and technology in principle afford. Some of the loudest sociologists and revisionist radicals want science to become something other than it is. That last group believes that indoctrination into their viewpoint is a legitimate aspect of teaching. That, together with other substantive differences, is a source of continuing tension among the disparate entities that have been lumped together recently under the umbrella of STS.

My Intellectual Dissatisfaction with Current STS

It is not obvious that STS should serve only the immediate purposes of those who brought it into being. The scientists who moved into policy-discussing did so in the belief that the wider society, indeed humanity as a whole, stands to benefit thereby. The academics who sensed the need for multidisciplinary work also believe that their endeavors to understand are necessary for the common public good. There is in fact consensus that STS should serve the purposes of the wider public; but as usual, there is dispute over what best serves the public interest. What is clear is that many of the current exercises of STS do not further our understanding of actual science in the actual world, in all the complexities and nuances that are of interest to the wider public including the media and the legislators and regulators.

Practicing scientists and technologists are neither last nor least among the potential audience for STS. I am not alone in having turned to science studies because my work within science led me to curiosity about metascientific matters. Why have I found STS frustrating? For one thing because, in John Ziman's words, "Unfortunately, the 'metascientists'—the historians, philosophers, sociologists, psychologists, economists and political scientists who describe and analyze science and technology from a variety of different points of view—have not yet come up with a coherent account of just how the research process actually works" (Ziman, 1994, p. 275).

When I was studying chemistry in college, in the late 1940s and early 1950s, I was already interested in "the big picture" or the *whole* picture; but there were not any history or philosophy of science courses available to me; and what I tried to read of philosophy of science turned me off. In the late 1960s and early 1970s my curiosity became pressing as I started to wonder why some very interesting topics were apparently ignored by science or treated as beyond the pale. So I came to read about attempts to demarcate science and pseudoscience, and that led to some history of science; and then I was lucky enough to become involved in Virginia Tech's Center for the Study of Science in Society and to learn from direct contact with such people as Joe Pitt, Larry Laudan, Rachel Laudan, David Lux, Arthur Donovan, Karin Knorr, and many, many others. I thought I had come into the right company, of very knowledgeable and insightful people who shared my interest in really understanding *everything* about how science works: how it has managed to bring us such wide-ranging, reliable, and deep understanding of Nature; how fallible and argumentative human beings could become such overachievers. I lapped up Thomas Kuhn and Robert Merton and John Ziman and a great deal more.

I joined 4S and attended some of its meetings, and felt rather at home. I am sure that practicing scientists and engineers were explicitly invited to join the Society in those early years. By the late 1970s, 16 percent of the membership described itself as physical or biological sciences or engineering and another 8 percent as administrative or general education (Nelkin, 1977).

But after a time, starting maybe a decade ago, I began to feel increasingly dissatisfied. There seemed to be a lack of coherence and progress in "science studies." It was not at all like chemistry, where interesting points are seized and built on and expanded and modified; and where the literature is so well organized and indexed that there is no excuse for ignoring important discoveries or for writing or saying something that has already been shown to be nonsense. One of the most fascinating papers I had come across was Bernard Barber's dense little "Resistance by Scientists to Scientific Discovery," published in *Science* in 1961. For years I looked for work that expanded those seminal insights, but without success. In the *Social Sciences Citation Index* as well as the *Science Citation Index* I periodically look for work building on it and continue to be disappointed.[9] I have had similar disappointment over Gunther Stent's notion of "premature discovery," which seems to have left barely a ripple. I recall what one of our sociologists said early in the life of our Center: that they were interested in science as a source of illustrative examples of social phenomena. But that is the very opposite of what I came to science studies for, namely for insights about what is *special* about science, what is *different* about science.

That is one of the circumstances, I believe, that easily turns scientists off STS. Most explicitly the "strong programmers" but to some extent all constructivists and relativists seem intent on showing how every aspect of scientific activity is significantly *the same* as what humans always do. So particulars and specifics are used to lead to always the same overgeneralizations—moreover the same ideological generalizations that the "analyst" brings to the case a *priori*. And it is true enough, of course, that all human beings are subject to emotions and self-interest, and *part* of what they do can be explained that way; but *only* part. What is *uniquely* interesting about *science* is how reliable knowledge has *nevertheless* been attained—so reliable that human beings of every stripe agree over it. So I appreciate the sociology of Merton and Zuckerman and Derek Price and Jonathan Cole and others who dig into the details of how science works in order to bring forth *unexpected, new* points of information or understanding. "[T]he whole aim of Mertonian analysis was (and is) to understand that which is socially and cognitively *distinct* about science compared with other forms of knowledge, such as religious beliefs or art, and which has given science such preeminence in modern society. This . . . is *the* question on which we build a sociology of science, but it is one which the relativists deny any warrant to, since they seek to deconstruct science's very claim to special authority and preeminence over other (non-scientific) forms of knowledge" (Gieryn, 1982). Scientists want *answers*, and *novelty*, stimulant to *more* and *new* studies. Harping always on human frailties of scientists not only misses the important points, and the big picture, it also amounts to denigration of science.

My intellectual dissatisfaction with STS has been joined by the perception that scientists and engineers are not—or are no longer—welcome. I hear it said that prominent people in science studies have asked, "What do we need scientists for?" That seems to be the implicit message when the Society for Social Studies of Science announces on the Internet that "4S includes (1) scholars in sociology, anthropology, history, philosophy, political science, economics, and psychology; (2) areas of study that fall outside of the traditional academic disciplines such as feminist studies and those addressing science and technology for the public; (3) studies of knowledge, policy, government, R&D, the uses of expertise, technological controversies, technology transfer, rhetorical and literary analyses, and studies of specific technologies." Grist for Gross and Levitt rather than a welcome to practitioners of science and technology.

As well as serving the wider public, STS surely has a responsibility to its own students. Particularly when graduate degrees are offered, some attention ought to be given to potential employment. It seems quite unlikely that many academic positions will open up in the foreseeable future. STS graduates might plausibly look for positions *mediating* between the technical and the nontechnical, policy-making communities—in research management, in assisting with environmental-impact considerations, in advising individual legislators and legislative groups, in science journalism. To mediate successfully calls for empathy, a "feeling for the organism" involved. Scientists and the public expect metascientists to have, if not deep knowledge then at least a genuine feeling for the organism that is scientific activity, in its complexity and diversity and nuanced distinctions.[10] To that end, STS should pay some heed to what the more reflective scientists have to say about science. Disrespect for any of the communities concerned is dysfunctional. STS graduates who instinctively denigrate science and technology are unlikely to find—or to keep for any length of time—employment funded by science and technology, which represents the bulk of what is likely to be available.

Disciplinary Dilemmas

My frustration with STS may stem in part from the *cultural* sorts of differences made notorious by C. P. Snow. My scientific background makes me want *progress* in knowledge and understanding but philosophers do not experience such progress in their own field and consequently do not look for it in philosophy of science or in STS. My chemistry background makes me lust for generalizations and interpretations that pull all the data together—but historians are trained to distrust and resist any such lust and will resist it in STS as in history per se. My disciplinary background makes me seek *decisive* ways to choose among possible viewpoints—but social and behavioral scientists inhabit "multiparadigmatic" fields in which the same data are disparately interpreted in the light of contrary preexisting theoretical stances, and they see neither need nor possibility for STS to acquire a single overarching paradigm. So what I look for in STS is not what many or most non-scientist STS practitioners look for.

These disciplinary distinctions explain, too, how some relativists can hold views that seem to scientists inane. Practitioners of the social sciences have not learned, in their own disciplines, much that is operationally indisputable, readily reproducible, and internationally agreed to; so they cannot easily conceive such a thing to be possible in *any* field. Knowing in their own discipline that ideology governs "knowledge" as well as theory, they presume that must be so in all fields.

That the subject matter of society is so much more complex than that of inert Nature is often said to be the reason why the social sciences are less rigorous (up to now at least) than the natural sciences. But this view is another instance of nonscientists lacking an understanding of science. The natural sciences deal successfully with immensely complicated systems incorporating numerous *simultaneously causal relationships*. In consequence, scientists see no reason of principle why STS should not be able to explain *everything* about science, and they are disappointed when STSers do not even dream of such a thing.

The disciplinary dilemma may be resolvable through recalling that STS should serve the wider public interest; and by recognizing that it is unlikely to thrive if it fails to do so. The wider public really does not care at all about disciplinary approaches or goals. It simply wants usable output from us. The communities of scientists and technologists, too, want useful explanations if not advice.

Surely STS has some knowledge to offer pertaining to what is likely to make research thrive and what is likely to throttle it; or regarding what to do as a society when there are hints or suspicions of risks to the ozone layer and of global warming and the like, about which definite, precise knowledge will not be available in time if those suspicions turn out to have been well founded. That is the sort of thing that the wider public would find useful from STS. In providing it, moreover, we would do well not to dwell on possible doubts about the validity of "naive" realism, for the society that funds our work and wants its fruits has no such doubts.

UNDER WHAT CIRCUMSTANCES IS STS LIKELY TO THRIVE?

Science itself, and technology, have long enjoyed high public status and prestige. STS, and the social sciences generally, have not. STS can thrive in the long run only if (1) the opulence academe has enjoyed since the 1950s continues or (2) STS demonstrates its usefulness to the wider public, including the technical community. (1) is already plainly not the case. (2) will not be served by doing as the radical wing of STS seems intent on—ignoring or seeking to denigrate or destroy the popular belief that science has produced a large body of reliable knowledge about how the world works and that it properly enjoys high prestige therefore. Or, as historian of science Stephen Brush formulated it in an article comparing earlier history of science to present scholarship in this field:

> The current fashion is to stress the social construction of scientific theories and concepts, and to deny that scientists are actually discovering the truth about the world, or that their efforts to do so have any moral or epistemological superiority to those of pseudoscientists, humanists, and theologians. The historian no

> longer assumes that scientific research is an admirable activity and scientific progress a benefit to society . . .
>
> Readers unfamiliar with recent publications in science studies may not believe that any reputable scholar actually holds such extreme views. (Brush, 1995)

If it comes to an overt confrontation between STS and science, science is bound to win. The scientific and technological communities continue to enjoy high, even unquestioned status and prestige among most of the population, who are clear what their standard of living depends on. Practitioners of science and technology are looked to not only for technical advice but also for *policy* advice regarding technical matters: thus Presidential Science Advisors are typically scientists or engineers, and the Nobel Peace Prize for 1995 was awarded to a physicist on behalf of the Pugwash Conferences on Science and World Affairs that he and other scientists had been principally responsible for organizing (Marshall, 1995; Selzer, 1995). Those communities also have the ear of the media, and themselves control important parts of the media. Among general magazines of science, *Scientific American* commands a circulation, in its English-language version, of 627,000.[11] The circulation of *Science* is about 155,000, that of *American Scientist* 103,000, of the *New Scientist* 101,000, *The Scientist* 50,000, *Nature* 31,000, *Issues in Science and Technology* 18,000, the *Bulletin of the Atomic Scientists* 13,000. By contrast, the circulation of *Science, Technology, and Human Values* is about 1,900, that of *Minerva* about 800, of *Social Epistemology* less than 200 (that of *Social Studies of Science* was not reported in Ulrich's, nor had it been in the previous edition).

Yet overt confrontation seems to have begun, with the publication of *Higher Superstition* in 1994. Much correspondence in *Chemical and Engineering News*, from some of the 135,000 members of the American Chemical Society, displayed approval of Gross and Levitt's polemic and dismay that their life's work is being denigrated, for example by the Smithsonian Institution.[12] What the Smithsonian did resulted in "cancellations to the museum's magazine (an important source of revenue), affected donations to other parts of the Smithsonian, prompted a letter-writing campaign to the Institution's new secretary, prompted that same secretary to delay work on several other potentially controversial projects, and threatened to affect the spring 1995 round of congressional hearings on the Smithsonian budget" (LaFollette, 1995). The American Physical Society joined the Chemical Society in protesting (Holden, 1994a). Earlier, there had been a storm over the proposed national standards for science education when a late draft "conveyed 'the really bizarre postmodern notion that somehow science is just a matter of social convention, rather than analysis of data' " (Culotta, 1994; see also Holden, 1994b).

One session of the General Meeting of the National Association of Scholars in November 1994 was devoted to the question of antagonism against science. The New York Academy of Sciences held a conference in June 1995, devoted entirely to that. A discussion of that meeting in *Chemical and Engineering News* concluded that "The deconstructionists and radical feminists and their collaborators on college campuses will fade away because their message is, at heart, ridiculous" (Baum, 1995). Much more was heard about "ridiculous" following the physicist Alan Sokal's "Experiment with cultural studies": Sokal published in *Social Text* a deliberate spoof of postmodernist discourse that the editors were unable to distinguish from the sort of article they publish; even though "about a half-dozen editors at the journal dealt with Professor Sokal's unsolicited manuscript" (Sokal, 1996a; Scott, 1996). He "structured the article around the silliest quotes about mathematics and physics from the most prominent academics, and . . . invented an argument praising them and linking them together . . . All this was very easy to carry off because my argument wasn't obliged to respect any standards of evidence or logic" (Sokal, 1996b, cited in Scott, 1996).

The scientific backlash will not soon fade away. An editorial in *Chemical and Engineering News*, "The Antiscience Cancer," recently acknowledged that scientists "should clean our own house and speak out when scientists overplay their findings or promise more than they can deliver. . . . However, if the mainstream scientific organizations, like ACS, the American Association for the Advancement of Science, the National Academy of Sciences, the Council on Chemical Research, and the International Union of Pure and Applied Chemistry just sit back and watch, the future of science, at least in the U.S., is bleak." "Presumably, tenure decisions and promotions at universities are based on scholarship, and academic scientists must take an interest in the academic decisions in other departments on campus. This is not a question of academic freedom, but rather one of competency. We should expose political correctness and fundamentalism that lead to misinformation about science."[3] That scarcely veiled threat to contest the academic position of those who attack science is worth noting most particularly because it comes not from some intemperate radical but from Allen Bard, editor of the leading ACS journal, one-time president of the ACS, who is known throughout the international community of chemists and scientists—as well as to this author personally—as a singularly judicious, level-headed person who has excelled in managerial matters calling for diplomacy as well as in technical tasks.

Scientists have come close to desperation as the ratio of research funds to researcher has steadily declined, and they are likely to vent their anxiety on any such obvious targets as those documented by Gross and Levitt and flushed into the open by Sokal. That the backlash did not come earlier may well have been due in part to the ignorance of most of the scientific community about the diatribes cited in *Higher Superstition*. Then too, scientists habitually defer to fellow specialists in other scientific fields and were therefore wont to defer also to philosophers, historians, sociologists, and others, presuming that they knew what they were doing in their specialized areas, even if that included venturing insights into scientific activity. But such deference wilts in the face of statements about science that scientists know to be absurdly wrong as well as insulting. I find it striking that John Ziman, who has long been so deliberately respectful toward social science, allowed himself in print the sign of impatience cited at the outset of this essay. Gerald Holton, again hardly an enemy of STS, has warned of the consequences when antiscience is incorporated into political movements (Holton, 1993).[13]

Organizational self-preservation of STS requires that it make its peace with science *on terms congenial to science and its publics*. STS must demonstrate that it has redeeming social value. I think that the *intellectual* self-preservation of STS requires the same.

NOTES

1. *Chemical and Engineering News*, 11 March 1996, p. 40.

2. I came to this through Kooks by Donna Kossy, Portland, OR: Feral House, p. 60.

3. For example, in *Travels into Several Remote Nations of the World* by Lemuel Gulliver, London: Benjamin Motte, 1726, particularly Part III—"A Voyage to Laputa, Balnibarbi, Luggnagg, Glubbdubrib, and Japan."

4. Papers by Thomas F. Gieryn and Elizabeth Hunt, and by P. G. Abir-Am at the 1987 meeting of the Society for Social Studies of Science, abstracted in *Science and Technology Studies*, 5:74–76, 1987.

5. In seminars at Virginia Polytechnic Institute and State University. In discussion, Shapin was unwilling even to restrict his assertion to the historical figures of several centuries ago whom he had actually studied. Pickering was also unwilling to concede that Nature constrains our science: "I hate that word 'constraints,' " he responded to my direct question.

6. "Vignettes: The Hierarchy of the Sciences," *Science*, 269 (11 August 1995):860, citing from N. Katherine Hayles in Michael E. Soule and Gary Lease (eds.), *Reinventing Nature? Responses to Postmodern Deconstruction*, WA: Island Press, 1995.

7. Among Bernal's books are *The Social Function of Science, Science for Peace*

and Socialism (with M. Cornforth), *Marx and Science, Science in History* (2 volumes, NY: Cameron Associates, 1954). See also Maurice Goldsmith, *Sage: A Life of J. D. Bernal* (London: Hutchinson, 1980). The approach and scope of this British stream of pre-STS can be gathered from the essays in Maurice Goldsmith and Alan Mackay (eds.), *The Science of Science* (Harmondsworth [Middlesex, U.K.]: Penguin, 1966 [rev. ed.]).

8. More than 150 programs and more than 2,300 courses in "Science, Technology and Society" were listed in Ezra D. Heitowit, *Science, Technology and Society: A Survey and Analysis of Academic Activities in the U.S.*, Ithaca, NY: Program on Science, Technology and Society, Cornell University, July 1977. Nearly 200 colleges and universities reported some teaching on ethical implications of science and technology—Ezra D. Heitowit and Janet Epstein, *Listing of Courses and Programs in the Field of Ethical and Human Value Implications of Science and Technology*, Ithaca, NY: Program on Science, Technology and Society, Cornell University, 1 January 1976.

9. Ron Westrum rightly pointed out to me in this connection that failure to cite important work may signify that the work has become so generally well known that citation is considered unnecessary. And it is true that some of Westrum's intriguing studies build on the theme of resistance to discovery and references therein). Yet Barber's paper makes some very specific suggestions as to possible reasons for resistance and circumstances in which it is likely to be found, and there seem not to have been attempts to explicitly test them.

10. A popular exemplar for those who would have science reformed if not entirely reinvented is Barbara McClintock, who according to Evelyn Fox Keller, insists that "one must have the time to look, the patience to 'hear what the material has to say to you,' the openness to 'let it come to you.' Above all, one must have 'a feeling for the organism.' " So too, I suggest, must STS have a feeling for the organism that is science if it is to understand science well enough to interpret it to others. In the following paraphrase from Keller (who in places quotes McClintock), I have merely substituted "science," "bit of science," or something similar for "plant" or the like, showing substitutions by underlined italics:

> One must understand how it grows, understand its parts . . . *science* isn't just a piece of plastic, it's something that is constantly being affected by the environment, constantly showing attributes or disabilities in its growth. You have to be aware of all of that . . . You need to know *science* well enough so that if anything changes, . . . you [can] look at it and right away you know what this damage you see is from . . .
>
> "No two *bits of science* are exactly alike. They're all different, and as a consequence, you have to know that difference," . . .
>
> This intimate knowledge, made possible by years of close association with *science* . . . is a prerequisite for . . . perspicacity.
>
> Good *STS* cannot proceed without a deep emotional investment on the part of the *STSer*. It is that emotional investment that provides the motivating force for the endless hours of intense, often grueling, labor . . .

> . . . reason—at least in the conventional sense of the word—is not by itself adequate to describe the vast complexity—even mystery—of *science. Scientific activity* has a life and order of *its* own that *STSers* can only partially fathom. No models we invent can begin to do full justice to the prodigious capacity of *science* to devise means for guaranteeing *its* own *continuing success.* On the contrary, 'anything you can think of you will find.' In comparison with the ingenuity of *science,* our *STS* intelligence seems pallid.
>
> . . . It is the overall organization, or orchestration, that enables *scientific investigation* to meet its needs, whatever they might be, in ways that never cease to surprise us . . .
>
> Our surprise is a measure of our tendency to underestimate the flexibility of *scientific work,* (Fox Keller, 1983, p. 198)

11. Circulation figures are from Ulrich's *International Periodicals Directory, 1994–1995,* 33rd ed., New Providence, NJ: R. R. Bowker.

12. For example, *Chemical and Engineering News,* 13 June 1994:4; 17 April 1995:5; 6 March 1995:4–5; 8 May 1995:4 and 63; 5 June 1995:4–5.

13. See the last chapter of Holton's *Science and Anti-Science.* Holton has carried the same message to the mentioned meetings of the National Association of Scholars (Boston, November 1994) and the New York Academy of Sciences (June 1995).

REFERENCES

Barber, B. 1961. "Resistance by scientists to scientific discovery." *Science* 134 (3479) September 1:596–602.

Bard, A. 1996. "The antiscience cancer." *Chemical and Engineering News* (22 April):5.

Barzun, J. 1994. "Psychotherapy Awry" (letter). *American Scholar* 63 (3) Summer:479.

Bauer, H. 1984a. *Beyond Velikovsky: The History of a Public Controversy.* Urbana: University of Illinois Press.

———. 1984b. "Velikovsky and Social Studies of Science." *4S Review* 2 (4) Winter:2–8.

———. 1986. "Frontier science and textbook science." *Science and Technology Studies* 4 (3–4):33–34.

———. 1986–87. "The Literature of Fringe Science." *Skeptical Inquirer* 11 (2) Winter:205–10.

———. 1992. *Scientific Literacy and the Myth of the Scientific Method.* Urbana: University of Illinois Press.

Baum, R. 1995. "Attacks on science require measured, reasoned response." *Chemical and Engineering News* (26 June):34, 85.

Brock, T. 1988. "Is it history? Philosophy? Or none of the above?" *The Scientist* (14 November):20.

Brush, S. 1978. *The Temperature of History.* New York: Burt Franklin.

———. 1995. "Scientists as Historians." *Osiris* 10:215–231.

Collins, H., and Pinch, T. 1993. *The Golem: What Everyone Should Know About Science*. Cambridge: Cambridge University Press.

Culotta, E. 1994. "Science Standards Near Finish Line." *Science* 265 (16 September):1648–1650.

Forman, P. 1995a. "Truth and Objectivity, Part 1: Irony." *Science* 269 (28 July):565–567.

———. 1995b. "Truth and Objectivity, Part 2: Trust." *Science* 269 (4 August):707–710.

Fox Keller, E. 1983. *A Feeling for the Organism: The Life and Work of Barbara McClintock*. New York: W. H. Freeman.

Freeman, D. 1983. *Margaret Mead and Samoa: The Making and Unmaking of an Anthropological Myth*. Cambridge, MA: Harvard University Press.

Gieryn, T. 1982. "Relativist/Constructivist Programs in the Sociology of Science, Redundance and Retreat." *Social Studies of Science* 12:279–297.

Greenstone, J. 1972. *The Messiah Idea in Jewish History*. Westport, CT: Greenwood (originally published 1906, Philadelphia: Jewish Publication Society of America).

Gross, P., and Levitt, N. 1994. *Higher Superstition: The Academic Left and Its Quarrels with Science*, Baltimore: Johns Hopkins University Press.

Heitowit, E., and Epstein, J. 1976. *Listing of Courses and Programs in the Field of Ethical and Human Value Implications of Science and Technology*. Ithaca, NY: Program on Science, Technology and Society, Cornell University, 1 January 1976.

Holden, C. 1994a. "Random Samples: History Slights Science." *Science* 266 (25 November):1327.

———. 1994b. "National Standards Finally Ready for Public Scrutiny." *Science* 266 (9 December):1637.

———. 1996. "Random Samples—History standards embrace science." *Science* 272 (19 April):351.

Kleinman, D. 1995. "Why Science and Scientists Are under Fire—and How the Profession Needs to Respond." *Chronicle of Higher Education* (29 September):B1, 2.

Kuhn, T. 1970. *The Structure of Scientific Revolutions*, 2n ed., enlarged. Chicago: University of Chicago Press.

LaFollette, M. 1995. "Editorial—Wielding History Like a Hammer." *Science Communication* 16 (3) March:235–241.

Holton, G. 1993. *Science and Anti-Science*. Cambridge, MA: Harvard University Press.

Marshall, E. 1995. "Physicist Wins Nobel Peace Prize." *Science* 270 (20 October):372.

Nelkin, D. 1977. *4S Review* 2 (1) Winter.

Nemecek, S. 1995. *Scientific American* (February):21, 22.

Ross, L. 1994. "Science in American Life, ACS-based exhibit opens at Smithsonian." *Chemical and Engineering News* (2 May):4, 5.

Scott, J. 1996. "Postmodern gravity deconstructed, slyly." *New York Times* (18 May):A1, 22.

Segerstråle, U. 1994. "Science by Worst Cases." *Science* 263 (11 February):837–838.

Seltzer, R. 1995. "Arms Control Work Wins Scientists Nobel Prize." *Chemical and*

Engineering News (23 October):9.

Snow, C. P. 1964. *The Two Cultures: And a Second Look*. New York: Mentor (original ed. Cambridge: Cambridge University Press, 1963).

Sokal, A. 1996a. "Transgressing the Boundaries: Toward a Transformative Hermeneutics of Quantum Gravity." *Social Text* (Spring/Summer).

———. 1996b. "A Physicist Experiments with Cultural Studies." *Lingua Franca* (May/June):62–64.

Soule, M., and Lease, G. (eds.). 1995. *Reinventing Nature? Responses to Postmodern Deconstruction*, Washington, D.C.: Island Press.

Spiegel-Rösing, I., and de Solla Price, D. (eds.). 1977. *Science, Technology and Society: A Cross-Disciplinary Perspective*. London: Sage.

Ulrich's International Periodicals Directory, 1994–1995, 33rd ed., New Providence, NJ: R. R. Bowker.

Westrum, R. 1982. "Social Intelligence About Hidden Events." *Knowledge: Creation, Diffusion, Utilization* 3 (3) March:381–400.

Wolpert, L. 1988. "A Sociologist Challenges the 'Validity of Science.' " *The Scientist* (14 November):20.

Ziman, J. 1994. *Prometheus Bound: Science in a Dynamic Steady State*. Cambridge: Cambridge University Press.

CHAPTER THREE

Some Patterns and Processes in the Development of a Scientific Sociology of Science

Notes from a Sixty-Year Memoir

BERNARD BARBER

EDITOR'S INTRODUCTION

In this chapter Bernard Barber, a pioneer of the field of sociology of science, shares with us some of his reflections on the changing relationship between science and science studies. As a field, the sociology of science was directly founded on the back of the scientific growth curve after the Second World War—more precisely, as part of the "Sputnik effect." During the following decades of postwar Mertonian sociology of science, the relationship between the sociologists and their objects of study was unproblematic.

Postwar sociology of science was also preoccupied with the relationship between science and society, particularly between science and democracy. One of the documents of this epoch is Barber's own 1952 book Science and the Social Order. *But Barber early on also planted a critical seed in science studies. In his his "Resistance by Scientists to Scientific Discovery," he demonstrated how various types of social judgments affect scientists' acceptance of one another's theories. (This* Science *article got over 500 reprint requests.)*

We may be rightly curious about the verdict of this founding father of the sociology of science when it comes to recent developments in this field. Barber acknowledges the importance of the "Strong Programme" in making also the content *of science a subject for sociological scrutiny. However, he is also critical, noting some weaknesses that have to be overcome before the sociology of science as a whole can move forward. These themes are further explored in his latest book,* Intellectual Pursuits.

WRITING THE HISTORY OF SOCIAL STUDIES OF SCIENCE

It is *past time* for us to begin to write the history of the development of the field of social studies of science. Such a history, of course, would be a sociological history. Indeed, such a history, since it would be the history of an emerging and ongoing science, would be much like all our other studies of science. We would be studying ourselves as scientists the way we study all other scientists. We would look for the patterns and processes, and for their various social structural and cultural determinants, of how and when our specialty emerged from near nonexistence into the complex and variegated maturing discipline it now is.

As a contribution to this history, I submit, in this chapter, a discussion of *some* of the patterns and processes in this development that I have experienced in my own work and observed in the work of our colleagues. It hardly needs be said that caution on the part of my audience is called for. This is an *initial* effort by *one* scholar; as such, it cries out for correction and amplification (Barber 1990). A satisfactory sociological history should be a continuing work-in-progress.

Here, then, are my provisional patterns, principles, and their determinants, for the development of the sociology of science.

THE APPLICATION OF GENERAL SOCIAL SYSTEM THEORY

In the marvelous sociological training I had received as an undergraduate and graduate student at Harvard during the thirties and forties from Sorokin, Merton, Parsons, and L. J. Henderson, I had been most attracted, as a result of Parsons's teachings and writings by general social system theory (Barber, 1993). No wonder, then, that I decided, in my first work after leaving Harvard, to show the usefulness of that theory and, concurrently, to improve and extend it. A colleague and I hatched a grandiose plan to do this, a plan that had two parts. The first part was to write a social system theory treatise, a text à la Samuelson in economics, laying out a social system analysis of society filled not only with that analysis but with as much comparative empirical work from sociology, history, and anthropology as we could find. The second part, such was the grandiosity of our plan, was to write a full volume on each of the social structural and cultural components of the social system that we had dealt with in only a chapter in our treatise. Both in the treatise and the special volumes we intended to include analytic and empirical discussions of social problems and social change.

Unfortunately, the collaboration with my colleague on the treatise was aborted for personal reasons and we agreed that he should write the treatise and I would write a book on one of the many social structural and cultural components of society. I chose to do a book on science, and the result was my *Science and the Social Order* (Barber 1952). Using social system theory, the book discusses the functions of rationality and science in the social system, the development of modern science, its social organization, science in the two authoritarian countries—Germany and Russia, the social process of discovery and invention, the processes of social control in science, and (as a harbinger of my future work on the ethics of scientific research on human subjects) the social responsibilities of science. I ended with an optimistic account of the nature and prospects of the social sciences, an account that I think has been justified by developments in the social sciences generally and not least of all in our own specialty.

Why did I choose science for this first book on each of the several social structural and cultural components of the social system?[1] I did so to illustrate the usefulness of social systems theory, not because I hoped to start a new specialty in the sociology of science. I chose science because it was the component I thought I knew the most about. As an undergraduate and graduate student I studied the work on the social aspects of science by Sorokin, Merton, Parsons, and Henderson. I did an undergraduate paper in an American Intellectual History course on Jefferson as a scientist. Especially in my graduate years, I came to know and study with the great historian of science, George Sarton, and to know the younger historians of science who were to play a great role in the remarkable development in the history of science that has occurred in the last forty years: people like I. Bernard Cohen, Henry Guerlac, and Giorgio de Santillana. I had been much impressed by the work on the nature and history of science by President James Bryant Conant of Harvard.

Despite Robert Merton's especially brilliant work on Puritanism and science and his pioneering essays on the norms of science, the sociology of science did not exist as a recognized and teachable subject. In the broader field of the sociology of knowledge, which we knew well particularly from Karl Mannheim, science was omitted as a privileged and independent category. Merton and I remained pretty much isolated characters. In my book, although scores of citations were provided, those cited were also mostly isolated scholars, not part of dense intercitation networks such as those we now know to be the clear sign of an active discipline.

THE POLITICAL AND IDEOLOGICAL RESPONSE

As we know from such recent cases as the so-called war on cancer and the practice of gene therapy, the development of scientific specialties is affected by the circumambient political and ideological environment. Among sociologists, my book was not especially welcome for at least two reasons. First, sociologists had not been trained in the natural sciences and felt uncomfortable with it. Second, social system theory was not in great favor because of the widespread hostility to Talcott Parsons's work and to functionalism more generally. The more general political and ideological environment also played its part. An American anti-Communist scientist who had written a book about what he called "the death of science in Russia" all but dismissed my book as the work of a misguided Marxist. By contrast, when the book was published in England, where there was a group of leftists much influenced by the so-called British scientific humanists like Bernal, Hogben, and Haldane, a reviewer said, Professor Barber, "though an American," seems to know the facts of reality. That was because in my analysis of Russian science I had pointed out what was favorable to science in the Russian social system as well as what was unfavorable. Incidentally, one American who admired my analysis of Russian science was an official of the CIA, not so identified explicitly, who invited me to come to Washington to set up a group to study Russian science. Those being McCarthy days and I having been an undergraduate member of the American Student Union, I had to tell him that I would not be a suitable government employee under the McCarthy rules.

The whole political and ideological environment in the United States changed, of course, with the Russian success with Sputnik in 1957. The American disdain for Russian science was abandoned and large resources were poured into the universities. The study of the history and social aspects of science was only one of the established and emerging scientific specialties in the natural and social sciences that profited hugely from this new and vast government financial support and approval. Because of the changed atmosphere, because of the establishment of large numbers of courses in "science and society," there was a new market for my book and two large paperback printings were issued that far outsold the earlier hard-cover edition.

THE INFLUENCE OF NONSYSTEMATIC THEORIES

Social system theory was not the only kind that I found useful in understanding the social aspects of science.

Various less systematically integrated pieces of theory were sometimes helpful. Because of his early theoretical work on the unanticipated consequences of social action, Robert Merton was fascinated by what he called "the serendipity pattern" in research, which occurs, as he said, when "the unanticipated, anomalous, and strategic datum exerts a pressure for initiating theory" (Merton, 1936; Merton, 1949, p. 99–101).

I was, of course, entirely familiar with this attractive piece of theorizing about the social processes of scientific discovery and, therefore, able to take full advantage of my knowledge when the opportunity offered. Such an opportunity did offer itself to my colleague, Professor Renee Fox, and me, when we learned, through a story in the *New York Times* and through personal acquaintances, that two distinguished medical research scientists at New York University and Cornell University Medical College, had both experienced, serendipitously, the "bizarre" and anomalous pattern of floppiness in test-rabbits' ears after the injection of the enzyme papain. One had gone on to make a minor medical discovery, the other, after some failed efforts, had gone on to others of his several research projects.

This seemed like a quasi-controlled experimental situation to us and we arranged long interviews with each of the two scientists to compare why one had succeeded and the other not. Both men had started out with erroneous preconceptions about cartilage, but the successful scientist, because he wanted to demonstrate the anomalous floppiness effect to his students, persevered. Eventually, through closely examining sections of the rabbits' ears, he saw that the cartilage was not rigid, as received knowledge had it. This discovery led on to his further work on cartilage. Not a great discovery, but typical of what happens frequently in scientific research.

We titled the paper we wrote about this work "The Case of the Floppy-Eared Rabbits: An Instance of Serendipity Gained and Serendipity Lost" (Barber & Fox, 1958). Perhaps partly because of its catchy title, this paper was widely noted and reprinted.[2]

NO THEORY AT ALL: THE IMPORTANCE OF A STRIKING EMPIRICAL FACT

Social science is not always theory-driven. Sometimes a massive or striking empirical fact generates research that then calls forth explanatory theory. This was the case for my work on the subject of resistance by scientists to scientific discovery.

In my wide reading in the history of science, especially in scientific biographies and memoirs, I came to see a recurrent pattern, one in which scientific discoveries were rejected for reasons apart from those of scientific substance or method. I decided to make a systematic inventory of examples of this pattern and look for some theoretical explanation. My research did turn up a number of examples that were new to me and I saw that such nonscientific reasons as philosophical preconceptions and status differences could explain the various cases of resistance. The most famous case, of course, was that of the resistance of von Nägeli to Mendel's great discovery.

At the invitation of one of the editors, I submitted my paper on this subject to *Science* magazine (Barber, 1961). Partly because of this prominent place of publication, and partly because my subject struck a nerve among many scientists, this paper was perhaps the most widely noted of all my many published papers. I received more than 500 requests for reprints, letters that often gave alleged personal examples, even personal visits, one from a critic of Einstein who claimed to have been resisted. Of course, for many of the tales I heard, I could give no judgment about the alleged resistance since I did not have the necessary details. But it was also clear that many of my readers did not get my point. I made it clear, I thought, in the article, that I was discussing only cases where initial rejection could be definitely attributed to specific sociological factors apart from legitimate substantive or methodological criticism. But in many cases, it seemed, the nerve I had struck was that connected with the difficulty many scientists have even with legitimate scientific criticism. What my experience in this cases showed was that, for all the power of the norms of objectivity and neutrality in science, working scientists put a great deal of passion into their work and resent even legitimate criticism.

THE IMPORTANCE OF NEW RESEARCH METHODS

In the 1970s some valuable research methods were newly introduced into the sociology of science, where they have flourished ever since. Using the focused interview method and the survey research method that had been strongly developed and emphasized at Columbia University's Bureau of Applied Social Research by Professors Paul Lazarsfeld and Robert Merton, my younger colleagues at Columbia, Jonathan and Stephen Cole and Harriet Zuckerman, brought new data and new analysis into the study of social stratification in science (Cole & Cole, 1973) and the origins, patterns of work, and problems of the scientific elite represented by the Nobel laureates (Zuckerman, 1975). The Coles also pioneered in the sophisticated use of citation analysis, a method of seeing various social patterns in science that, despite recognized weaknesses, has become a valuable research tool for the sociology of science. The sociology of science has a large debt to Eugene Garfield's Institute of Scientific Information, which provided the data for citation analysis and for much else besides in the processes of effective communication among scientists.

I was sufficiently attracted by these new methods that I resolved to use them myself in some new research on the ethics of medical research on human subjects. With a group of able younger colleagues, John J. Lally, Julia Loughlin Makarushka, and Daniel Sullivan, all well trained in the new methods, I published *Research on Human Subjects: Problems of Social*

Control in Medical Experimentation (Barber, 1973; Barber, 1990, part III), which reported the results of several large studies of these matters. I will discuss this and related work further, below.

In this section I must also pay tribute to pioneering, innovative quantitative work on the social processes of science by Professor Derek J. deSolla Price. Derek Price taught us all how indispensable quantitative data could be in our field.

THE WEAKNESS OF "THE STRONG PROGRAM"

Another major and exciting movement that occurred in the sociology of science beginning in the 1970s and that has lasted to the present is the movement that labelled itself "the strong program." This was a movement that came out of Great Britain, particularly out of Edinburgh, but that has spread to Germany, Holland, France, and also the United States.

It has had several features that attracted my admiration. First, in the writings of David Bloor, a professional philosopher who took up a Wittgensteinian position, it has had a strong philosophical base, albeit one that was ultimately relativist. Second, it has been carried forward by the research and writing of an exceptionally able and energetic group of social scientists: Barry Barnes, David Edge, Michael Mulkay, Harry Collins, Trevor Pinch, Karin Knorr-Cetina, Wiebe Bijker, G. N. Gilbert, Steven Shapin, Donald MacKenzie, Bruno Latour, Steven Woolgar, and others. Third, these scholars made intensive studies not only of the social organization of science but of the substantive ideas of science. This overcame a shortcoming I had felt in my own work. Not trained in science, I had to limit myself, with a few exceptions like the resistance paper, to studies primarily of its internal social organization and its place in the larger social system of society. These new people had, in some cases, like David Edge, formerly been practicing scientists. The others worked seriously and intensively at understanding the substance of scientific ideas. Finally, I admired the energy of this group: they organized programs to train students, they started and have edited successfully what is the premier sociology of science journal (*Social Studies of Science*), they called conferences and were diligent attenders and presenters of papers about their work at all relevant meetings, and they organized new groups to discuss their work. Every scientific specialty eventually needs some committed members to do all these tasks, and "the strong program" group took them up with energy and resulting success.

For various reasons, however, early on I came to feel that these strengths of "the strong program" went along with overriding weaknesses. First, it had a strong strain of ontological relativism about science; in many cases, their work seemed to deny the possibility of an objective, cumulative science. And presumably, since they stressed the principle of reflexivity, this meant that their own work had no objective scientific standing. Second, as a sociologist, I found unsatisfactory their usual attribution of the determinants of scientific work to the broad and vague concept of "interests." Insofar as "interests" were specified, they seemed to be only political and economic ones, not interests stemming from strongly held values or from the cognitive concerns of working scientists. Implicit in this exclusive emphasis on political and economic interests, I came to feel, were definite antiauthoritarian values and ideologies. Paradoxically, social scientists with strong antiestablishment values were denying the importance of values and norms.

In sum, I believe that "the strong program" group has strengthened the conviction among us sociologists of science that not only the organization but also the substantive ideas of science are in part determined by a whole variety of social structural, cultural, and personality factors. But it is clear that their tendency to relativism and their inadequate sociology will not do to carry this important premise of all our work forward.[3]

THE CROSS-FERTILIZATION OF SOCIOLOGICAL SPECIALTIES

The sociology of science has been enriched by its cross-fertilization with other sociological specialties. As a result of my early and continued experience with the groundbreaking teaching and writing of Talcott Parsons in the sociologies of medicine and the professions, I had myself made some contributions in those fields. But I was struck, especially in the sociology of medicine as much as in the sociology of science itself, with the lack of attention to the substantive scientific data of the field. As a result of some work I did in collaboration with Professor Renee Fox at a pharmaceutical firm, I became interested in the sociology of drugs and went so far as to make up a tentative table of contents for a book on that subject. In one of my many visits to the stimulating informal lunch-seminars held regularly by Bert Brim, then the president of the Russell Sage Foundation, he said, most presciently, that the subject of drugs was going to be one of the most important in the future. When I told him about my book outline, he offered to support my writing it. I eventually accepted his offer, and the final product brought in the substantive science as well as the problems of organization in the drug field (Barber, 1967).

In addition to such topics as discovery and testing processes and education and communications processes, I discussed the functions and problems of the professional specialists who did research on new drugs. One of the problems I discussed was the questionable ethics of using human subjects in some of that research. Mine was a pioneering treatment of that problem, but because of a couple of national scandals, the use of human subjects without their informed consent and without due consideration of a satisfactory risk-benefit became a hot political and social topic. In my attempt to analyze the problem from a social science point of view and collect sound data on the extent and sources of the problem, I was given further support by the Russell Sage Foundation. One result, mentioned above, was what we called the "Two Institution Study" published in the book *Research on Human Subjects* (Barber, 1977). Our key explanatory hypothesis was what I called "the dilemma of science and therapy," and the findings from our survey and network data analyses, showed the strength of that hypothesis.

Another result of this work was that, for some years, I became active in the public and political events aroused by the concern over the use of human subjects. I testified in U.S. Senate hearings, I became a member of committees drafting new rules for professionals, I addressed many professional groups, I was asked to check some foreign research projects on reproductive biology that had been subsidized by the Ford Foundation, and I became a member of an Institutional Review Board at a medical research institution, the kind of review board that is now mandated for all research institutions using human subjects. In all of this activity, thus, I was playing my small part in the development of what I consider a valuable moral improvement in the conduct of medical scientific research.

THE CROSS-FERTILIZATION OF DISCIPLINES

Another valuable enrichment of the sociology of science has come from its cross-fertilization with the history of science and the philosophy of science. All three of these disciplines were very small and mostly separate when my *Science and the Social Order* was published forty-odd years ago. All three have not only grown considerably since then but have become increasingly intertwined. I have already mentioned the importance of David Bloor's philosophical writings for "the strong program." Perhaps the most influential book for our field that brought together the history, philosophy, and sociology of science was Thomas Kuhn's *The Structure of Scientific Revolutions* (Kuhn, 1962).

The cross-fertilization of these three disciplines has not only served to show the special and different contribution that each has to make to our understanding of science but has improved each of them by showing both their limitations as against the others but also their necessary interactions (see Barber, 1990). For example, much of the history of science now uses sociological theory to explain what it formerly only described, and sociologists have learned to avoid imperfect popular history and to use and even occasionally create accounts that meet professional historical standards. On their part, philosophers of science have profited from knowing the best of the new history and sociology of science.

THE IMPORTANCE OF VALUES AND IDEOLOGIES

Perhaps the most striking example in the recent sociology of science of the importance of values and their associated ideologies, perhaps the most direct critique of "the strong program's" assertion that only political and economic interests affect the development of scientific ideas, is the vitality and burgeoning of the feminist movement in our field. This movement has raised questions about such important matters as whether women scientists create different kinds of scientific ideas than men do, about their careers, and about their productivity. The sociology of science has been enlarged and stimulated by this movement (Zuckerman et al., 1991).

IN BRIEF CONCLUSION

How do I sum up? How do I account briefly for my long journey in the sociology of science and for its present state? In a phrase, *This is not where I came in.* The sociology of science has made enormous progress in the last forty-odd years and is now a relatively mature and flourishing specialty. It has achieved this state as the result of a set of diverse, multiple, and interactive theoretical and other social structural, cultural, and personality factors. Political and economic interests are only a few of these important determinants. We do not need to be ontological relativists about science and its development. Science is an essential functional component of the culture of all societies and has its own degree of autonomy as well as its dependence on all the other functional components of the social system. On these philosophical and sociological premises, we can go forward to a continuing theoretical, moral, and practical success.

NOTES

1. As a further step in the plan for social system theory, in 1957, I published *Social Stratification: A Comparative Analysis of Structure and Process.*

2. For some other examples of the valuable results of pieces of theory, see Merton (1961) and Merton (1968).

3. For a powerful critique of the assumptions and substance of the work of "the strong program" see Schmaus et al. (1992). A very important additional reference in connection with "the strong program's" weaknesses is Cole (1992). Cole's book, I feel, offers the best way forward for the sociology of science. Finally, an absolutely indispensable discussion on the subject of "truth and objectivity" is the two-part review by Paul Forman in *Science* (Forman, 1995a; Forman, 1995b).

REFERENCES

Barber, Bernard. 1952. *Science and the Social Order*. Glencoe, Ill.: The Free Press.

———. 1957. *Social Stratification: A Comparative Analysis of Structure and Process*. New York: Harcourt Brace.

———. 1961. "Resistance by Scientists to Scientific Discovery." *Science* 134:596–602.

———. 1967. *Drugs and Society*. New York: Russell Sage Foundation.

———. 1975. *Research on Human Subjects: Problems of Social Control in Medical Experimentation*. New York: Russell Sage Foundation.

———. 1987. "Big Science." *Isis* 78 (294):589–591.

———. 1990. *Social Studies of Science*, New Brunswick, N.J.: Transaction Books.

———. 1993. *Constructing the Social System*. New Brunswick, N.J.: Transaction Books.

———, and Renee Fox. 1958. "The Case of the Floppy-Eared Rabbits: An Instance of Serendipity Gained and Serendipity Lost." *American Journal of Sociology* 64:126–136.

Cole, Jonathan R., and Stephen Cole. 1973. *Social Stratification in Science*. Chicago: University of Chicago Press.

Cole, Stephen. 1992. *Making Science: Between Nature and Society*. Cambridge, MA.: Harvard University Press.

Forman, Paul. 1995a. "Truth and Objectivity. Part I: Irony." *Science* 269:565–567.

———. 1995b. "Truth and Objectivity. Part II: Trust." *Science* 269:707–710.

Kuhn, Thomas. 1962. *The Structure of Scientific Revolutions*. Chicago: The University of Chicago Press.

Merton, Robert K. 1936. "The Unanticipated Consequences of Purposive Social Action." *American Sociological Review* 1: 894–904.

———. 1949. *Social Theory and Social Structure*. Glencoe, Ill.: Free Press.

———. 1961. "Singletons and Multiples in Scientific Discovery." *American Philosophical Society, Proceedings* 105:470–486.

———. 1968. "The Matthew Effect in Science: The Reward and Communications Systems of Science." *Science* 199:55–63.

Schmaus, Warren, Segerstråle, Ullica, and Douglas Jesseph. 1992. "The Hard Program in the Sociology of Scientific Knowledge. A Manifesto." *Social Epistemology* 6 (3):243–265.

Zuckerman, Harriet. 1975. *Scientific Elite: Nobel Laureates in the United States*. New York: Free Press.

———, Cole, Jonathan R., and John T. Bruer (eds.). 1991. *The Outer Circle: Women in the Scientific Community*. New York: W. W. Norton.

CHAPTER FOUR

Anti-Antiscience: A Phenomenon in Search of an Explanation

Part I. Anatomy of Recent "Antiscience" Allegations

ULLICA SEGERSTRÅLE

EDITOR'S INTRODUCTION

Just as some scientific activists have attempted to explain what they see as an "antiscience phenomenon" among critical analysts of science, so a sociologist may be intrigued by this very "anti-antiscience phenomenon" itself, and try to provide some explanation for it, in turn. This chapter explores some possible reasons for the recent attack on "the academic left" by warriors against "antiscience." The author notes that "antiscience" appears to be something of a constructed category itself. To understand what may be driving the pro-science activists in the Science Wars, she concentrates on their cognitive and moral/political commitments, although she later considers also practical and strategical motivations. (In her book Defenders of the Truth: The Battle for Science in the Sociobiology Controversy and Beyond, *the author further argues for a continuity between some of the fundamental issues underlying the sociobiology controversy and the Science Wars.)*

In search of typical concerns of those who are worried about "antiscience," the author compares and contrasts the allegations of "antiscience" in the Science Wars with two other recent situations where scientists have accused others of antiscience, both taking place in the 1980s. In the first episode in the United States, it was natural scientists*—rather than humanists and social scientists—that were accused of antiscience. In the second episode, taking place in Britain, two physicists in* Nature *blamed Karl Popper's philosophy of science for the decline in the national science budget! Their worries about spreading antiscience sentiments were remarkably similar to those of Gross and Levitt. Gross and Levitt, however, refused to engage in "moral archeology" of this sort. Instead they faulted their targets for distorting someone like Kuhn.*

TAKING THE SCIENCE WARRIORS SERIOUSLY

In the mid-1990s a number of academics in various fields found themselves accused of "antiscience". Sometimes there were even dark references to an antiscience movement. As such, of course, opposition to science is nothing new; antiscience sentiments have been characteristic of many historical periods (e.g., Brush, 1978; Holton, 1993). What I am interested in here, however, are the specific claims about antiscience by the recent proscience activists. Exactly what did "antiscience" mean for them, and what prompted them to such fierce action against purported promoters of antiscience?

The relevant anti-antiscience manifestos are three: Gerald Holton's *Science and Anti-Science* (1993), Paul Gross and Norman Levitt's *Higher Superstition: The Academic Left and Its Quarrels with Science* (Gross & Levitt, 1994), and *The Flight from Science and Reason* (Gross, Levitt, & Lewis, 1996), the edited proceedings of a conference organized by the National Association of Scholars.[1] A somewhat less visible, but equally passionate forum for warnings about antiscience has been the popular journal *The Skeptical Inquirer*. Common for all recent anti-antiscience warriors is their contention that an antiscience threat exists and that widespread antiscience attitudes will have dangerous social consequences.

There is, undoubtedly, a certain paradox in the whole idea of crying "antiscience!" Because, presumably, for this cry to have any effect, there would have to be a big enough audience on whose outrage against antiscience one can count. But if such an audience exists, this would seem to imply that sufficient support for science does exist, too, which in turn suggests that there was no serious need for a rallying cry, after all. However, I note this only in passing. I am less interested in contrasting the actual situation with the rhetorical strategies of the anti-antiscience warriors than in examining the broader concerns prompting their recent attack.

Higher Superstition attracted considerable interest in American academia. Some saw the book as "the scientists" finally striking back at those academic colleagues who had already for quite some time been enthusiastically analyzing and criticizing science from the outside, as it were. But most scientists had, in fact, to be told that an antiscience threat even existed. Gross and Levitt themselves—after they had learned about their existence—had clearly become infuriated with the recent constructivist and relativist (or "postmodern") challenges to science.[2] However, in pointing out that the real danger of the cultural left's "hostility" to science was not to science itself, but rather to culture as a whole, Gross and Levitt appealed not only to scientists, but also to a larger group of academics who had already been sensitized to the issues in the ongoing "Culture

Wars" on American university campuses (on the Culture Wars see, e.g., Searle, 1990; Berman, 1992; Gitlin, 1995).[3]

What ensued was the Science Wars, one more of those polarizations so typical in contemporary academia, with loose coalitions on both sides.[4] One one side were the proscience activists joined by academic traditionalists from different fields, on the other the different groups that Gross and Levitt collectively accused of "antiscience," and conveniently united under umbrella terms such as "cultural constructivism" or "postmodernism."[5] In Gross and Levitt's hands, postmodern (and other radical) humanists in this way became rather artificially connected with the social constructivists from the field of sociology of scientific knowledge. The intellectual roots of these two academic groups are in fact quite different and they had relatively little to do with each other—at least before the start of the Science Wars.[6]

Such seemingly uninformed misclassifications have made it easy to ironically dismiss the anti-antiscience warriors' view as nothing but a social construction itself (cf. Martin, 1996). As meta-analysts of science, Gross and Levitt did appear as outsiders, who applied their own "external" standards regarding such matters as who was and was not a social constructivist, what were the actual aims of social studies of science, and so on. This often resulted in views that were demonstrably not shared even by the sternest critics of social constructivism within STS.[7] Finally, because politically the proscience warriors' counterattack against current science critics appeared to fit so clearly in with a conservative social agenda, it seemed to prove the point of those who perceived an intimate connection between science and social power. For the latter, what was going on in the Science Wars was that science, connected to powerful social interests, was striking back against the academic left's devastating, politically progressive criticism of it.

It may, therefore, be tempting to dismiss or explain away the views of the current anti-antiscience warriors as simply motivated by politically conservative interests. For a sociologist, however, it is more interesting to take these new academic activists seriously. What exactly were they after? Why did they talk so much about science and reason? Why did they regard almost any type of criticism of science as "antiscience"? And while these proscience fighters may have been wrong when they lumped together various types of science critics under a common umbrella and called them "hostile" and "ignorant of science," the very fact that they did this opens up one more window to the world of their minds. In the following, I will try to understand and contextualize the views of the most vocal proscience activists in order to locate the exact nature of their concern.

I will start by comparing and contrasting the recent Science Wars with two other recent episodes where scientists also cried "antiscience." The first involves the nature-nurture controversies in the 1970s and 1980s. There the charges of "antiscience," although similar to the present ones, were not directed against social scientists or humanists but against bona fide *natural scientists* instead! In the Science Wars we do not typically see such attacks.[8] The second precedent goes back to 1980s Britain. In 1987, two physicists accused philosophy of science for spreading "antitheses" about science and thereby undermining science funding in that country. The exact reasoning of those scientists can give insights into the *potential* logic guiding the later American anti-antiscience warriors. However, the latter followed their British colleagues only part of the way. As we shall see, Gross and Levitt's militancy against "antiscience" and persistent emphasis on "science and reason" may be better understood by examining the particular scientific worldview that they themselves can be expected to have absorbed: the post–World War II idea of science as an important *cultural* institution. (I will return to this topic in Part II, chapter 5.)

Political allegiances do not seem to have much explanatory power when it comes to the activities of the anti-antiscience warriors. Although many have regarded the Science Wars as a simple opposition between Right and Left, the political distinctions have turned out to be much more subtle. Despite the misleading subtitle "the academic left" in Gross and Levitt's book, it is becoming increasingly clear that the biggest political conflict existed between different factions *within* the contemporary academic left itself. I will defer the discussion of this problem, too, to Part II, together with my suggestion that an important hidden problem in this conflict was the tendency of both sides to equate "reason" with "science."

Finally, a caveat is in order. This chapter's focus on the cognitive commitments of the proscience fighters does not, of course, mean to ignore the existence of various types of *strategical* considerations as well. The same thing is clearly true also for the targets of the anti-antiscience warriors, as I will briefly discuss in Part II. Whatever the case, there needs to be no contradiction at all between cognitive commitments and strategical considerations when it comes to the activities of scientists. In fact, scientists typically seem to be pursuing both at the same time, which is why scientists are best characterized as "optimizers" (Segerstråle 2000, chapter 15).

THE VARIETIES OF "ANTISCIENCE"

In *Higher Superstition* Gross and Levitt very clearly specified the target of their attack, a fact that was not lost on the media. For instance, in the fall of 1994, *The Chronicle of Higher Education* dramatically quoted the authors as dismissing "the relativism of the social constructivists, the sophomoric skepticism of the postmodernists, the incipient Lysenkoism of feminist critics, the millenialism [*sic*] of the radical environmentalists, the radical chauvinism of the Afrocentrists" as "unscientific and antiscientific nonsense" and as a "bizarre war against scientific thought and practice being waged by the various ideological strands of the academic left" (Cordes, 1994; Gross & Levitt, 1994, pp. 252–253). Indeed, even the superficial reader could hardly miss the authors' hit list on the book's early pages: the Marxist view of science as a tool of capitalism (either as such or "refurbished as the doctrine of 'cultural constructivism' "), the radical feminist conception of science as affected by gender bias, the multiculturalist condemnation of Western science as inherently inaccurate, and radical environmentalist views holding science responsible for ecological disaster (Gross & Levitt, 1994, pp. 4–5).

In a later article entitled "Antiscience in Academia—Knocking Science for Fun and Profit," Gross and Levitt discussed what they believed to be the guiding motives of this "academic" or "cultural" left and how the latter's approach differed from "serious" criticisms of science. According to the authors, the reason for the popularity of the new postmodern critique of science was that it did not require anyone to master science—it was a criticism for lazy academics, informed by a particular "cultural left" ideology, deeply hostile to science. The worry was that this ideology was spreading:

> A new and fashionable cottage industry has appeared among the intelligentsia, especially among academics. Its principal activity is to issue quantities of arrogant and hostile criticism of science. The specific content of the examined science rarely comes into it, for the simple reason that *the trendier critics don't bother to study the science seriously* . . .
>
> What concerns us, however, is the new brand of criticism . . . whose underlying motive is . . . the ideology of what has been called the "academic" or "cultural" left. . . .
>
> Its view of science is today highly influential among the academy . . . and increasingly so outside the campus gates. In part, this is due to a tactical exploit of the new science critics: *they have avoided the professional scrutiny of scientists and others devoted to reason as a means of understanding the world.* . . . Most important is the debasement of the public discourse of science and technology, a discourse already inadequate to the complexity of global issues at whose heart lie scientific questions. (Gross & Levitt, 1995, italics added)

In their mission against antiscience, Gross and Levitt found natural allies in the people behind *The Skeptical Inquirer*, the journal of the Committee for the Scientific Investigation of Claims of the Paranormal (CSICOP), an organization devoted to critical inquiry into pseudoscience and paranormal claims, Gross and Levitt's above-mentioned article appeared in the March/April 1995 issue of this journal, but already one year earlier (Spring 1994), a whole issue was devoted to "The Antiscience Threat." In that issue, Paul Kurtz, the founding chairman of CSICOP, identified the following ten reasons for what he believed were examples of antiscience sentiments in contemporary society:

1. The anxiety about a possible nuclear holocaust after World War II
2. Fears generated by the environmental movement
3. Widespread phobia about chemical additives
4. Suspicion of biogenetic engineering
5. Widespread attack on orthodox medicine
6. The growing opposition to psychiatry
7. The phenomenal growth in "alternative health cures."
8. The impact of Asian mysticism in the form of Yoga and various spiritual cures
9. The revival of fundamentalist religion even in advanced scientific and educational societies; creationism in the United States
10. The growth of multicultural and feminist critiques of science education

The same 1994 issue also contained excerpts from Harvard physicist and historian of science Gerald Holton's then brand-new book *Science and Anti-Science* (Holton, 1993). In his book, he made a direct connection between antiscience sentiments in society and potentially disastrous consequences for the political system as a whole: "[H]istory has shown repeatedly that a disaffection with science and its view of the world can turn into a rage that links up with far more sinister movements." Moreover,

> History records that the serious and dedicated portion of the anti-science phenomenon, when married to political power, does signal a major cultural challenge. At its current level, this challenge may not be an irreplaceable threat to the modern worldview as such. But it cannot be dismissed as just a distasteful annoyance either, nor only as a reminder of the failure of educators. On the contrary, the record from Ancient Greece to Fascist Germany and Stalin's U.S.S.R. to our day shows that *movements*

> *to delegitimate conventional science are ever present and ready to put themselves at the service of other forces that wish to bend the course of civilization their way*—for example, by the glorification of populism, folk belief, and violence, by mystification, and by an ideology that arouses rabid ethnic and nationalistic passions. . . . (Holton, 1994, italics added; excerpted from Holton, 1993, Chapter 6[9]).

Holton identified four prominent sections among the current "cohort of delegitimators" of science:

1. A type of modern philosopher who asserts that science is nothing more than a social myth, or sociologists of science (such as Bruno Latour) who wish to abolish the distinction between science and fiction
2. A small group of alienated intellectuals, who have chosen to attack science (such as Arthur Koestler)
3. The resurgence of "Dionysians"—people with New Age type preferences and a profound opposition to the idea of objective data; such preferences are traceable to the 1960's counterculture and nineteenth-century Romanticism
4. Radical feminists such as Sandra Harding, who calls for a total "intellectual, moral, social, and political revolution" against androcentrism and faith in the progressiveness of scientific rationality (Holton, 1993, pp. 153–154).

Thus, as we see, a number of different things have recently been called "antiscience." The common concern, however, for those who worried about "the antiscience phenomenon" was that they saw the contemporary critique of science criticism as fundamentally *irrational*. Abandoning science and reason would be socially dangerous, they warned, because we need them for public discourse and for solving global problems. Meanwhile, the anti-antiscience warriors did not enter into any discussion about the nature of science itself. Against various types of alleged "superstition" or "ideology" was upheld a picture of science that appeared largely unproblematic. In this respect, the writers on antiscience appeared to have a similar conception of science as the *The Skeptical Inquirer*, in which the picture of science has always been relatively idealized, and probably much "harder" than many practising scientists would be willing to admit.[10]

But there was also genuine puzzlement. Gross and Levitt said they found enigmatic the academic left's "open hostility to the *actual content* of scientific knowledge and toward the assumption, which one might have supposed universal among educated people, that scientific knowledge is reasonably reliable and rests on a sound methodology" (Gross & Levitt, 1994, p. 2). They perceived this hostility not only as a denial of the Enlightenment idea of progress, but also as a direct identification of science with ideology—all this among academics who regarded themselves as progressive:

> There is something medieval about it, in spite of the hypermodern language in which it is nowadays couched. It seems to represent a rejection of the strongest heritage of the Enlightenment. It seems to mock the idea that, on the whole, a civilization is capable of progressing from ignorance to insight. . . . All the more shocking is the fact that the challenge comes from a quarter that views itself as fearlessly progressive—the veritable cutting edge of the cultural future. . . . These critics of science do not repine for the traditional mores and devout certainties of a prescientific age. They accuse science itself of a reactionary obscurantism , and they revile it as an ideological prop of the present order, which many of them despise and hope to abolish. (Gross & Levitt, 1994, p. 3)

This grouping together different types of critique and analysis of science under the same umbrella term "hostility to science" may not have correctly characterized the attitudes of many of the current science critics. At least some of those attacked by Gross and Levitt would have retorted—and did retort—that they in fact *liked* science, but that science had been misunderstood, needlessly mystified, and oversold, and that their aim was simply to present science as it really was to help the public understanding of science.[11] Others, again, did fit the anti-antiscience warriors' description, particularly if "hostility" implied that science was seen as a symbol of power and as supporting the social status quo. Some recommended abandoning objectivity and going for different "standpoint" epistemologies instead (e.g., Harding, 1991; for an analysis of standpoint epistemologies, see chapter 8 in this volume); still others believed that science needed to include a multitude of different voices in order to attain the best possible objectivity (e.g., Longino, 1990). What was important here, however, was the fact that the proscience fighters *perceived* the many different types of current criticisms of science as motivated by the same major sentiment of "hostility" to science—that is, by "antiscience."

Let us now take a look at the last time allegations about "antiscience" were made in American academia. As we shall see, the anti-antiscience statements by leading proscience activists at that time sound strikingly similar to those we have heard in the Science Wars—with one major exception. In the 1980s, the accusation of "antiscience" was directed against natural scientists instead of humanists and social scientists. Let us see what insights in the reasoning of the current anti-antiscience warriors such a comparison can bring.

CAN SCIENTISTS BE ANTISCIENCE? THE LINK BETWEEN THE SCIENCE WARS AND RECENT NATURE-NURTURE DEBATES

In the Science Wars, "antiscience" was used to denote efforts by outsiders to science, or by members of *other* academic fields than natural science. Antiscience advocates were typically described not only as hostile but also as ignorant of science. In *Higher Superstition*, Gross and Levitt's targets were a motley bunch. They included Stanley Aronowitz, Bruno Latour, Steven Shapin, and Simon Schaffer as representatives for "cultural constructivism," Steven Best, Andrew Ross, and N. Katherine Hayles as standard bearers for postmodernism. Sandra Harding, Evelyn Fox Keller, Helen Longino, and Donna Haraway were the chosen feminist theorists, and Carolyn Merchant, Dave Foreman, and Jeremy Rifkin stood for radical environmentalism. Also included were movements that Gross and Levitt believed had been affected by the antiscientific rhetoric of the academic left: AIDS activism, animal rights, and Afrocentrism (Gross & Levitt, 1994, p. 14).

As their choice of Latour shows, Gross and Levitt did not quite succeed in picking the most representative people for their critique of "cultural constructivism" (their category that combined social constructivism and humanist science criticism). For instance, although they made much of constructivism in their book, they largely missed what the science studies community *itself* would have considered leading social constructivists, while including Aronowitz, a writer and popularizer rather than science studies theorist himself.[12] But then again, "cultural constructivism" was Gross and Levitt's own category. The criterion for inclusion on Gross and Levitt's list appeared to be how well a person fitted certain criteria of "*language and intellectual temperament*" that the authors believed characterized the academic left. Another criterion was popularity, particularly among undergraduates.

But others believed that scientists, too, could be antiscience, and that there were some scientists within the academic left that Gross and Levitt had actually missed in their book. In his review of *Higher Superstition* the philosopher Michael Ruse took issue with Gross and Levitt about exactly this matter (Ruse, 1995). Ruse asked why the authors in their criticism of the "academic left," did not include such highly visible left-wing scientists like Stephen J. Gould and Richard Lewontin. As political critics of science, these activists had persistently attacked research on the biological foundations of human behavior (sociobiology, behavioral genetics, and the like); moreover, their views were popular among humanists and social scientists.

Interestingly, in a separate context, also Edward O. Wilson, the author of *Sociobiology* (Wilson, 1975) echoed the same sentiment that important left-wing scientists were missing. Wilson, however, did not criticize Gross and Levitt directly for excluding Gould and Lewontin—instead, he tried to compensate for this lack himself. In his autobiography *Naturalist*, published in the Fall of 1994, Wilson addressed the storm around sociobiology, in which some Harvard colleagues, notably Lewontin and Gould, were highly visible. A reader may wonder, though, why Wilson insisted on calling the critics of sociobiology *"postmodern,"* of all things? It was certainly not a term he had used earlier in his rebuttals of his critics, and the use of "postmodern" seemed anachronistic when used to refer to academic left-wing criticism in the 1970s and early 1980s.

However, seeing Wilson's glowing endorsement of *Higher Superstition* on its back cover, it is not hard to imagine what had happened. Gross and Levitt's persistent usage of terms like 'ideology' and 'the academic left' only too painfully reminded Wilson of his own experiences with left-wing academics. In the sociobiology controversy, they had viciously attacked what he regarded as serious science as a piece of "bourgeois ideology" designed to support social power holders (see, e.g., Segerstråle, 2000, chapter 2). In fact, the case of Wilson shows that the vague expression "the academic left" had considerable *heuristic power*. It enabled Wilson to reframe the sociobiology controversy to fit the current Science Wars, casting his opponents as early postmodern critics and himself as an early proscience warrior. In this way, his old opponents, could now be retroactively classified as proponents of antiscience. (This was, in fact, the gist of Wilson's talk about science and ideology at the first conference of the National Association of Scholars conference in Boston in November 1994; Wilson, 1995.)

Wilson's position was, in fact, connected to an earlier episode where scientists were declared "antiscience." We go now to the eminent Harvard microbiologist and proscience activist, Bernard Davis, Wilson's late colleague and supporter. In the 1970s, Davis had been one of the staunchest warriors against critics of IQ research and behavioral genetics; later on, he became a supporter also of sociobiology. Davis saw himself as a defender of science against antiscience within science itself, tirelessly writing essays and editorials, collected in 1986 as *Storm over Biology* (Davis, 1986). In his Foreword to the book, Edward Shils, the editor of *Minerva*, even warned about an antiscience movement led by scientists:

> The antiscience movement that has grown up in recent years is ignored by most of the scientific community. But it derives much of its force from the support of a small group of scientists and it may accomplish real mischief. . . .
>
> The disaffected scientists charge that the application of scientific knowledge is often pernicious, and some of its branches are even inherently dangerous. Sometimes coupled with this criticism is the view that modern science is driven by a desire to dominate both man and nature, and that it therefore seeks alliance with the earthly powers in the polity and the economy. Another frequent argument is that those who practice scientific research are members and supporters of "the elite," disregarding the welfare of the populace, excluding them from deliberations and decisions, and choosing problems without concern for the needs of ordinary persons. A more specific accusation is that certain branches believe in the inequality of human beings and attempt to demonstrate it.
>
> Some critics further assert that scientific knowledge is in fact only a part of the ideology of bourgeois society. . . . its pretensions to objectivity are alleged to be baseless. The detachment that has long been a source of pride for scientists would then be simply a pleasing fiction. An important school of the sociological study of science takes the position . . . that practical interests and "social position" enter into the very categories and the criteria of validity of knowledge. (Shils, 1986, italics added)

This quote from the 1980s addresses many of the concerns of the 1990s' critics of antiscience, too. Toward the end of the quote, we in fact see an early attempt to connect the social constructivist program in the sociology of science to an antiscience stance. (Shils was a Mertonian-type sociologist of science himself.) What is striking, however, is that, for Shils, these widely different criticisms of science together formed something that he called "the antiscience movement," and that he believed that it was *scientists* who were here in charge. In other words, Shils's and Davis's claims about antiscience in the 1980s were in fact *stronger* than those of the 1990s' anti-antiscience activists, who were "merely" attacking humanists and social scientists.

Although Shils went on to say that "only a handful of those who swim in the currents of antiscience do so equally in all of them," he still suggested that "these various currents do represent a *coherent body of beliefs*" (my emphasis). Shils and Davis, thus, were concerned with what they believed to be *a "syndrome" of antiscience attitudes.* Interestingly, though, unlike the 1990s' proscience warriors who seemed preoccupied with such abstract things as science and reason, Shils and Davis spelled out some very *practical* concerns about the effects that antiscience sentiments might have on science itself. They worried that if this type of antiscience thinking spread, it would negatively affect the public support of science and the recruitment of scientists, and even undermine the scientists' morale (Davis, 1986; Shils, 1986).

From the context of Davis's book it is clear that what Shils had in mind when he talked about "a small group of scientists" was those scientists vocally opposed to such things as sociobiology, behavioral genetics, and mental testing. For Shils, as for Davis, the mere fact that these scientists were *critical* of research in these fields was proof enough that they were against *science* as such. But were these scientific opponents to research in heredity and behavior really *anti*science, or were they merely *critical* of science? If one scrutinizes the people typically pointed to—Gould and Lewontin—one can find many instances where these purported antiscientists themselves in fact were speaking up *for* science. For example, Gould had on occasion said things such as: "As a practicing scientist, I share the credo of my colleagues: I believe that a factual reality exists and that science, though often in an obtuse and erratic manner, can learn about it" (Gould, 1981). But it is true that these critics had indeed protested against what they considered *"bad" science*—which for them was exactly those fields that people like Davis, Shils, and Wilson had been enthusiastic about, particularly sociobiology and mental testing. (For a discussion of these matters and clashing conceptions of "good science" in general, see Segerstråle, 2000.)

Thus, Gould was not a credible antiscientist then or later: he was rather against bad science. Lewontin, too, was consistently preoccupied with bad science, which he regarded as ideologically inspired, scientifically false, *and* socially dangerous. And when it comes to social constructivism at least of the Strong Programme type, Lewontin is on record as being explicitly opposed to it.[13]

What can be concluded from the above? In their opposition to antiscience or "bad" science, *both* the anti-antiscience warriors and their targets—the scientists accused of antiscience—were worrying about the *same* thing: keeping science pure and free from ideology. It is only that they all meant different things with "ideology." For Gross and Levitt, "ideology" was the whole set of views among the "cultural left"; for Wilson and Davis it was left-wing (particularly Marxist) ideology threatening objective scholarship. For the critics of sociobiology and related fields, again, "ideology" was bourgeois ideology aimed at buttressing a social belief in innate inequality (see, e.g., Lewontin, Rose, & Kamin, 1984; Levins & Lewontin, 1985).

At this point it becomes especially interesting to go back to Michael Ruse's charge that the authors of *Higher Superstition* did not criticize Gould and Lewontin for being antiscientists, even though the latter were notorious political critics of science. It is hard to tell what Gross and Levitt thought about Lewontin, because he did not appear in their book at all. However, Gould did figure quite prominently, and Gross and Levitt had nothing but praise for him. In fact, Gould was presented as a model for "incisive" historical and cultural criticism of science. Indeed, for Gross and Levitt, Gould—although leftist—was an *acceptable* critic of science, and was explicitly contrasted with unacceptable ones, such as the constructivists (Gross & Levitt, 1994, p. 56).

But wait a minute! Did not Bernard Davis, the late comrade-in-arms of the current proscience fighters, single out exactly Gould and his IQ critique in *The Mismeasure of Man* (Gould, 1981) as an example of "antiscience"—even worse, of "Neo-Lysenkoism"—in a long review article in the early 1980s (Davis, 1983)?[14] So how come in the 1990s *Mismeasure*, the very same Gould book that inspired Davis to write his "Neo-Lysenkoism" article, was warmly recommended by Gross and Levitt for its "analytical sanity" (Gross & Levitt, 1994, p. 263)? At the very least one can conclude that within the anti-antiscience position there has been serious disagreement over time as to who and what exactly should be included under the label "antiscience," and that the political lines have been quite unclear. This again reinforces the suspicion that "antiscience" is not a "natural kind" but rather a category with changing content—at each particular time constructed for a particular purpose by scientific activists.

KARL POPPER'S ADVERSE EFFECT ON THE BRITISH SCIENCE BUDGET: A THESIS IN 1980S BRITAIN

It is not hard to understand that committed scientists may become possessed by a holy rage when science seems to them unfairly attacked. But obviously the media have something to do with this, too. Science criticism has been turning into a form of entertainment—scandals of any kind do sell.[15] Books from the 1980s and early 1990s had titles such as *Betrayers of the Truth* (Broad & Wade, 1983), *False Prophets* (Kohn, 1986), *Impure Science* (Bell, 1992), and in 1991 the cover of *Time* magazine (26 August) was devoted to "Science under Siege." Older cases in the history of science were reexamined and relabeled and used as "canned" case studies in articles and books about fraud (cf. Segerstråle, 1995). Later, the turn came to electronic mail and to e-mail mailing lists such as SCIFRAUD. (The following piece of information may actually be taken as an indication of the seriousness with which the warriors against antiscience did take their mission. At one point, Gross reportedly took on himself to personally rebut the incoming messages on the SCIFRAUD mailing list. The journal *Lingua Franca* quoted Gross as saying: "I started off as a lurker. But these idiots pissed me off so much, I just had to respond." After that, Gross was said to respond "daily," refuting "every paranoid charge," Zalewski, 1995.)

Still, there is the question about the empirical relationship between purported antiscience sentiments and the real situation for science. The objective situation for science may have nothing to do with purported antiscience attitudes, even if they were widespread in the general public (which, incidentally, it seems they were not around the time of publication of Gross and Levitt's book).[16] Many have suggested that budget cuts and other problems were mainly produced by such larger factors as the tapering off of the growth curve of science (e.g., Ziman, 1994), or the end of the Cold War and its exaggerated science budget. Scientists themselves realized that, as a result, many assumptions about science, developed during the unprecedented growth of science after the Second World War, might have to be completely rethought (Byerly & Pielke, 1995; see also Cholakov's and Ziman's chapters in this volume).

A further puzzle is, of course: Why did Gross and Levitt concentrate exclusively on antiscience in *academia*? Why did they not, like Holton and Kurtz, address "obvious" cases of antiscience sentiments in the general public instead? Good candidates would have been religious fundamentalism or New Age beliefs. According to polls in the 1990s, about half of the United States population were believers in some version of creationism.[17] Meanwhile, scientific literacy was alarmingly low (e.g., Lawler, 1996)—a situation made worse by the media, whose sensationalist reporting often blurred the distinction between fact and fiction. This was clearly recognized as a problem by the scientific community. In 1996 *Science* reported on a Sunday evening television broadcast called "The Mysterious Origins of Mankind" (National Broadcasting Corporation, 25 February, 1996), a program that presented the creationist argument about the simultaneous existence of humans and dinosaurs on Earth in a factual way—and further insinuated that the scientific establishment was suppressing the evidence (Holden, 1996).[18]

A possible explanation for Gross and Levitt's focus on academia could be that they belonged to a special type of scientists that I have elsewhere called "weeders" (Segerstråle, 2000, chapter 11). Weeders believe that "bad" knowledge products ought to be weeded out at the source—that is, inside academia—before such things get out to the innocent public. In this respect, Gross and Levitt would indeed be similar to the before-mentioned critics of sociobiology and IQ testing, who believed in debunking "bad" science (or to Ned Feder and Walter Stewart, the two scientists at the

National Institutes of Health, who took on themselves the unusual task of weeding out errors and fraud in science (e.g., Segerstråle, 1993).

It should be noted, however, that although Gross and Levitt and others have held the "academic left" responsible for spreading antiscience sentiments, they have not explicitly accused these academics for closing down the Supercollider, or blamed them for cut-downs in the science budget. But, incredible as it may sound, something like this was in fact argued a decade ago in Great Britain. In 1987 in a lengthy commentary in *Nature,* two physicists, T. Theocharis and M. Psimopoulous, did indeed declare that the decreasing funding for British science was a direct consequence of the prevailing negative attitudes among academic philosophers of science! So who was the main culprit? Gross and Levitt had at least recognized Paul Feyerabend as "one of the thinkers directly responsible for initiating the chain of ideas leading to the cultural constructivist view of science," and Thomas Kuhn as "the most often cited" (p. 49), even though they ended up focusing their blame on the present-day "academic left." In contrast, the British physicists decided to get at the original culprit. For them, the root of all evil was Karl Popper!

In the 1980s, the situation for science in Britain was much the same as the 1990s' one in the United States; the cut in science funding just happened there somewhat earlier. The *Nature* piece by Theocharis and Psimopoulos was a clear bid in a larger ongoing discussion about proposed drastic changes in British universities and funding of research. Surprisingly, these physicists started out by a type of self-criticism: they suggested that the bad situation was really due to the lack of vigilance on the part of the scientific community! They saw themselves as conveying an important message to the world:

> Public spending on science has declined in other countries, too. But a combination of reasons peculiar to Britain has made the situation perhaps the worst in any advanced industrial state. . . . [O]ur objective is to identify and endeavour to combat what we consider to be the most fundamental, and yet the least recognized cause of the present predicament of science, not only in Britain but throughout the world. . . . In this article we argue that the British scientific and philosophical communities, and in particular the RS [Royal Society] have to their cost neglected one important factor in implementing their policies, and that therefore the financial crisis is to a considerable extent self-inflicted. (Theocharis & Psimopoulos, 1987)

What was this factor that the scientific community had neglected at its peril? The answer was: the general spread of antiscience sentiments, or, as Theocharis and Psimopoulos formulated it, of "epistemological antitheses." They now went on to mention several recent BBC television programs and popular articles devoted to themes such as "Science . . . Fiction?" and "The Fallacy of Scientific Objectivity," all reaching the same conclusion, namely that "[t]he gradual recognition of these arguments may affect the practice, the *funding* and the institutions of science." And what was the result of these programs and articles? A decline in the public spending on science! As Theocharis and Psimopoulos told the readers of *Nature*: "At least in Britain, the repercussions of these mistaken arguments are already happening. Scientists in other countries be duly forwarned" (Theocharis & Psimopoulos, 1987).

One piece of evidence that the two physicists held up as a proof of a direct link between declining government funding and a climate of doubt about science was a statement made already in 1971 by a Parliament member, who later (1976–1979) became the Secretary of State for Education and Science. She had issued the following warning:

> For the scientists, the party is over. . . . Until fairly recently no one has asked any awkward questions . . . Yet there is a growing suspicion about scientists and their discoveries. . . . It is within this drastically altered climate that the dramatic decline in expenditure on scientific research in Britain is taking place, (Shirley Williams, 1971, quoted in Theocharis & Psimopoulos, 1987)

What might this "drastically altered climate" conceivably refer to? 1971 sounds rather early for identifying a "growing suspicion" about science—at least one caused by social constructivists or postmodernists. At that time, the social constructivists were just discovering people like Kuhn and Feyerabend and largely still engaged in a rear battle against Mertonian norms. What occupied the social sciences at this time was, rather, a debate about values and ideology in social research. But within *philosophy* there did exist a vigorous discussion about scientific epistemology, objectivity, rationality, and the like. In this general discussion about the nature of scientific reasoning typically participated also some scientists, like Nobel laureate Peter Medawar, who contributed such spirited popular articles as the famous "Is the Scientific Paper a Fraud?" (Medawar, 1963), and a long discussion on the contrast between traditional statements about method and what scientists really do (Medawar, 1969). And, of course, this was the time of James Watson's publication of *The Double Helix* (Watson, 1968), which provided a surprising backstage glimpse of creative scientific work. It seems, then, that if anything, scientists and philosophers of science felt that the social position of science was strong enough to afford public speculation about it. There was no doubt that science demonstrably worked—the only question was how.

What is interesting is the clear structural parallel between the mid-1980's initiative by Theocharis and Psimopoulos and Gross and Levitt's actions less than a decade later. Both these sets of scientists wished to alert their "innocent" fellow scientists as to what was "really" going on. However, the British duo went one step further than their later American anti-antiscience colleagues: They directly criticized the Royal Society and other scientific bodies for lack of vigilance and for not having responded to media programs propagating "epistemological antitheses" about science. They also admonished the scientists who had participated in these programs for ignoring the fact that such programs could contribute to a climate of opinion that might threaten not only science but also society.

Still, the greatest difference was that the British physicists saw the decreasing public funding for science as a *direct result of the academic discussion about the nature of scientific epistemology.* Let us follow their reasoning to see how they reached this conclusion. The British duo had borrowed the idea of epistemological antitheses from the philosopher David Stove. (Stove had also labeled Popper an "irrationalist"; Stove, 1982.) They went on to list the antitheses as including "skepticism, agnosticism, 'criticismism' [a term invented by Stove], cynicism, relativism, anarchism, nihilism." Interestingly, just like Gross and Levitt, also Theocharis and Psimopoulos snidely remarked on the widespread popularity of such antiscience attitudes ("the endorsement of these antitheses saves one from the painstaking effort of discovering new truths"). And just like Gross and Levitt, they also warned that the antitheses had potentially serious consequences for science and society. But here the British duo became much more specific than their American colleagues. They boldly declared the existence of a *direct causal connection*, as follows:

> i) Intellectual bankruptcy entails financial bankruptcy.
> ii) Epistemological anarchism entails social anarchism.
> iii) Epistemological relativism, 'criticsmism' and nihilism entail scientific chaos, confusion and stagnation. (Theocharis & Psimopoulos, 1987)

But what was the actual *mechanism* through which such "isms" might conceivably produce a cut in the science budget? The authors realized that this had to be made clear. According to them, "the logical steps leading from the antitheses to the inescapable conclusion that the funding of science should be cut are not usually spelled out in detail. We spell them out now." Their logic went as follows:

> In the golden age of science, it was believed that verified theories of science were true and everlasting. But with the increasing acceptance of the antitheses, these exalted old ambitions of science are seen as bogus. According to epistemological relativism, science should no longer claim superiority for its method and the knowledge that it produces. . . . Scientific theories are now considered to be temporary and dispensable. Furthermore, by denying truth and reality, the antitheses reduce science to a pointless, if entertaining game; a meaningless, if exacting exercise; and a destinationless, if enjoyable journey . . .
>
> *Having lost their monopoly in the production of knowledge, scientists have also lost their privileged status in society*. . . . It is the duty of those who want to save science, both intellectually and financially, to refute the antitheses, and to reassure their paymasters and the public that the genuine theories of science are of permanent value, and that scientific research does have a concrete and positive and useful aim—to discover truth and establish reality. (Theocharis & Psimopoulos, 1987, italics added)

As we see, Theocharis and Psimopoulos at the very least provided a detailed model of the *possible* connections between academic discourse and financial support for science, something that had been left largely implicit in the current American discussion. They also formulated a course of action to defend science: to refute the "antitheses" and declare the positive value of science. And that part of their message was shared by the later American proscience warriors, too. In fact, Gross and Levitt can be seen as directly picking up on Theocharis and Psimopoulos's call for scientists to respond to the antitheses. Others, like Lewis Wolpert, have taken the other possible tack of reassuring the paymasters and the public that all is well with science.[19]

As we saw, in their "logical" search for the real culprit, the British physicists went all the way back to Popper. What was the rationale for this? For them (following Stove), it was Popper who had originally abandoned the idea of verifiability in favor of falsifiability and who, with his conjectures and refutations, had introduced the notion of *scientific truth as temporary*. From there it was all the way downhill to Kuhn, who abandoned the idea of truth in favor of transient paradigms, and finally to Feyerabend, who declared that science had no other foundation than the principle "anything goes!" This may be so, but it is odd that in 1987 Theocharis and Psimopoulos did not attack the then quite vocal social constructivists and relativists who were actually actively spreading the "antitheses" in academia and popular media. For instance, Harry Collins was one of the contributors to the "Science . . . Fiction" program on BBC (Collins, 1987).

It is even more remarkable that these physicists should have upheld their own causal *model*, making Popper the villain, against the well-known *fact* that Popper himself saw a positive social role for his epistemology of falsification (e.g., Popper, 1976).[20] To their credit, many readers of *Nature* protested against Theocharis and Psimopoulos's attack on Popper as unfair. And, not surprisingly, Popper himself was quite unhappy with the developments in philosophy, history, and sociology of science. In an interview in the *Scientific American* a couple of years before his death, he described the overall situation in the field of science studies in very negative terms. In fact, he described social constructivism as the hubris of sociologists who wanted to present themselves as more important than the natural scientists (Horgan, 1992).

Let us now return to Gross and Levitt. In contrast to Theocharis and Psimopoulos, who believed in tracing and morally condemning the person ultimately responsible, Gross and Levitt appeared quite willing to exculpate the theorists of science "who started it all." For instance, even if they mentioned Kuhn as the most quoted authority among the cultural left, they noted that Kuhn's work had "so often been vulgarized and distorted by the cultural constructivist school" (p. 56). In fact, they suggested that it took a special reading to misunderstand Kuhn's message so badly! As a defense of Kuhn, they invoke the fact that he himself in his reply to his critics in the second edition of his book (Kuhn, 1970) denied that his book lent support to relativism (p. 139).[21] But if Kuhn's intent was so obvious, how come a whole philosophical symposium on Criticism and the Growth of Knowledge was devoted to trying to figure out exactly what Kuhn was saying (Lakatos & Musgrave, 1970), including a paper identifying twenty-three different uses of the term 'paradigm' (Masterman, 1970)? Gross and Levitt also told their readers that Feyerabend "now expresses deep reservations about the outcomes of this line of thought" (p. 49). The basis for this assertion are some formulations in a brief 1992 article of Feyerabend's, entitled "Atoms and Consciousness." But again, this was not the Feyerabend most scientists or other people knew (for one of the last interviews with Feyerabend, see Horgan, 1993).

We see, then, that the American anti-antiscience warriors refused to engage in the Theocharis and Psimopoulos's type of "moral archeology," which would presumably have led them to Kuhn and Feyerabend (and perhaps even Popper) as the individuals who inspired later developments in cultural criticisms of science and the sociology of scientific knowledge. This refusal to identify Kuhn's influence is even more remarkable, since this is clearly at odds with the beliefs of the social constructivists themselves, who do point to just such an inspiration from Kuhn (independently of what Kuhn himself may or may not have meant) (e.g., Collins, 1983). Gross and Levitt seemed to believe that the meaning of Kuhn's book was somehow fixed once and for all, and that others did not have the right to "incorrectly" interpret it! With this type of move, at the same time, it seems that they wished to *reclaim Kuhn for science*, away from his unfortunate "kidnapping" by the humanities and social sciences.[22] Other candidates for culprits might have been French poststructuralists and postmodernists, such as Foucault and Lyotard, who undoubtedly "started it all" for the postmodernists. But Gross and Levitt left also them alone. Unlike their British colleagues, Gross and Levitt were quite uninterested in "logical" derivation of "true" culprits—they preferred to concentrate on the most well-known contemporary expositors of science criticicism.[23]

NOTES

1. The National Association of Scholars in 1994 and 1995 took the lead in alerting the educated public about the threat of antiscience, especially by arranging well-publicized conferences. The first one of these, "Objectivity and Truth in the Natural Sciences, the Social Sciences, and the Humanities" was held in Boston in November 1994. The second one, "The Flight From Science and Reason," was held in New York in June 1995. Its edited proceeedings were published first by the New York Academy of Sciences and later issued as a paperback by Johns Hopkins University Press under the name of *The Flight from Science and Reason* (Gross, Levitt, & Lewis, 1996).

2. Paul Gross, a biologist, was the then director of the Center for Advanced Studies at the University of Virginia; Norman Levitt is a professor of mathematics at Rutgers University. Gross has since retired. Both had attended university seminar series where postmodern and constructivist approaches to science were taken for granted.

3. This explains the attendance and contributions to these conferences not only by scientists but also by, for example, historians. For a representative mix of academics, see the contributors to *The Flight from Science and Reason*.

4. The term the "Science Wars" was coined by the editor Andrew Ross for a special issue of *Social Text*, which was intended to be a collection of criticisms of *Higher Superstition*. That was the very issue that also contained Alan Sokal's unsolicited contribution (Sokal 1996a). While Sokal's hoax deflected the message of that special issue, the expanded book version, *The Science Wars*, restored the editor's original intent (Ross, 1996).

5. The latest umbrella term seems to be "STS" (Gross, 1997). This is, however, an obvious misapplication of the term "STS." See Bauer's and Fuller's chapters for a history of STS.

6. Traditionally, there has been little historical connection between these groups (cf. Fuller's chapter). Gross and Levitt seem to have felt that they could

collapse these two groups on the grounds that "they publish in the same journals" (Gross & Levitt, 1994, p. 81). The overlap is in reality small, at least in regard to mainstream social studies of science. Another matter is that the anti-antiscience warriors themselves may have now encouraged more cooperation between the humanists and social studies scholars.

7. See, for example, Schmaus, Segerstråle, and Jesseph (1992); Cole (1992). Note particularly the *epistemological* nature of this critique.

8. An exception is zoologist Edward O. Wilson, a friend of Gross and Levitt's, who has indeed tried to pin the label "antiscience" on his Harvard biology colleague Richard Lewontin (Wilson, 1995). This in turn can be explained by the fact that Wilson's current anti-antiscience stance is a carryover from the 1970s and 1980s' sociobiology and IQ wars, where his late friend Bernard Davis declared criticism of research in the biological foundations of human behavior "antiscience." This episode is dealt with later in this chapter.

9. Holton also carried this message to the two National Association of Scholars conferences in 1994 and 1995.

10. To its credit, *The Skeptical Inquirer* does try to keep an open mind about paranormal phenomena, but sometimes the writing is militantly black and white and the CSICOP conferences are veritable celebrations of science.

11. Harry Collins and Trevor Pinch are saying this, even though others have perceived particularly their *The Golem: What Everybody Should Know about Science* (Collins & Pinch, 1993) as an attempt to underplay the capability of science to produce reliable knowledge, and therefore as potentially detrimental to public debate (Segerstråle, 1994; Turney, 1993). Meanwhile Steve Fuller (1995) explains that many sociologists of scientific knowledge see themselves as *helping* scientists with the public understanding of science. And Latour (1995), in response to Gross and Levitt's allegations, asserts that science studies practitioners do like science and technology very much and that "the spirit of confrontation and debunking is totally foreign to most of us." He also states that the "militant scientists and philosophers" ought to see science studies practitioners as their real allies in defending science against the rest of the world, since they can help "elucidate the material and social conditions that enable it to grow and expand."

12. An explanation for this could be that they are largely concerned with authors that are popular in "cultural studies" and "feminist studies" courses in undergraduate curricula. This could be why, for instance, they cite Stanley Aronowitz's presentation of David Bloor and the Strong Programme, rather than Strong Programme authors themselves (pp. 52–53). Indeed, they seem to believe that "Aronowitz represents cultural constructivism with all its philosophical and political cards on the table, so to speak. . . . Other theorists and publicists of the constructivist schools are more circumspect in their claims and cagier in their tactics" (p. 56). Again, the reason they dwell particularly on Latour is probably his widespread current popularity with humanists. However, Latour does not see himself as a social constructivist and has for some time been challenging his colleagues in the Strong Programme and the sociology of scientific knowledge (e.g., Latour, 1992, 1993, 1995; Latour & Callon, 1992; Collins & Yearley, 1992a,b; Bloor, 1997).

13. Together with two other critics of "biological determinism" in *Not in Our Genes* (Lewontin, Rose, & Kamin, 1984), Lewontin is on record as explicitly voicing his *disapproval* of the Strong Programme in the sociology of scientific knowledge. Just like the current critics of antiscience, Lewontin *also* believes in the existence of a "true" science. Indeed, it was just because they believed in true science that the authors of *Not in Our Genes* likened themselves to a fire brigade, who had to put out now this, now that, dangerous piece of scientific nonsense (Lewontin, Rose, & Kamin, 1984, p. 266).

14. With "Neo-Lysenkoism" Davis meant that people like Gould, Lewontin, and others were influenced by political (Marxist) ideology in their science. He also tried to demonstrate that Gould was not only a Marxist, but also (because of that) a bad scientist. Gould responded indignantly to this charge (Gould, 1984).

15. For instance, for analyses of the scandal around the Nobel laureate David Baltimore, see Segerstråle (1993), Kevles (1998).

16. The results from several studies from the mid-1990s indicated that contrary to the belief of some, the American public did trust science, despite the fact that it was rather ignorant of science (Lawler, 1996, Woolley, 1995).

17. Numbers (1992, p. ix), quoted in Hollinger (1995, p. 453). Later, in the edited conference volume *The Flight from Science and Reason* (Gross, Levitt, & Lewis, 1996), Gross does indeed more explicitly recognize the problem of creationism.

18. In regard to creationism, Gould has been doing a valuable one-man job over the years, defending science/evolutionism against antiscience/creationism.

19. See a discussion of Lewis Wolpert in Chapter 1.

20. Indeed, as Popper said in his autobiography (Popper, 1976), he was deliberately looking for an epistemology that when applied to the social realm would have politically desirable consequences. Thus, his ideas of conjectures and refutations, and falsification have their direct parallels in democratic ideals as against totalitarian regimes.

21. They might have added that Kuhn himself was ever since deeply unhappy about the unintended consequences of his ideas and tried to retreat to a more conventional-sounding position. After all, *The Structure of Scientific Revolutions* initially appeared as a contribution to the *International Encyclopaedia of Unified Science*, Otto Neurath's grand logical positivist effort.

22. The reason they regard Kuhn as someone who should be defended could have to do with their sympathy for Kuhn's original mission, which was after all, to help Harvard president John Conant in his post–World War II promotion of science through new, popular courses in the history of science (cf. Westman, 1995). See the next chapter for a discussion of the cultural status of postwar science.

23. It was instead Alan Sokal together with Jean Bricmont who picked up the idea of going back all the way to the French postmodernists and challenging them. In their new book, *Impostures Intellectuelles*, Sokal and Bricmont (1997) take most respected postmodern French intellectuals to pieces. However, according to one reviewer, the weak point of the book is that they have "cherry-picked the worst excesses from the philosophers' admittedly clumsy forays into science, ignoring the large body of their work which is worthwhile." Not surprisingly, this has created resentment in France (Boulet-Gercourt, 1997, see also Callon, 1999). The book was translated into English in 1998 (see chapter 1, note 18).

REFERENCES

Bell, R. 1992. *Impure Science: Fraud, Compromise, and Political Influence in Scientific Research*. New York: John Wiley.

Berman, P. (ed.) (1992). *Debating P.C.* New York: Bantam Doubleday Dell.

Bloor, D. (1997). Remember the Strong Program? *Science, Technology and Human Values* 22 (3):373–385.

Boulet-Gercourt, P. 1997. "In Defense of the Emperor's New Clothes." (Review of Alan Sokal and Jean Bricmont: *Impostures Intellectuelles*, Paris: Odile Jacob), *The European* 9–15 October: 58.

Broad, W., and Wade, N. 1983. *Betrayers of the Truth: Fraud and Deceit in the Halls of Science*. New York: Simon and Schuster.

Brush, S. 1978. *The Temperature of History.* New York: Burt Franklin.

Byerly, R., and Pielke, R. 1995. "The Changing Ecology of United States' Science." *Science* 269 (15 September):1531–1532.

Callon, M. 1999. "Whose Imposture?" Physicists at War with the Third Person. *Social Studies of Science* 29:261–86.

Callon, M. and Latour, B. 1992. "Don't Throw the Baby Out with the Bath School!" Pp. 343–368 in Pickering, A. (ed.), *Science as Practice and Culture*. Chicago: University of Chicago Press.

Collins, H. 1983. "The Sociology of Scientific Knowledge." *Annual Review of Sociology* 9:265–285.

———. 1987. Letter to the Editor. *Nature* 330 (24/31 December 1987):689–690.

———, and Pinch, T. 1993. *The Golem: What Everybody Should Know about Science*. Cambridge: Cambridge University Press.

———, and Yearley, S. 1992a. "Epistemological Chicken." Pp. 301–326 in Pickering, A. (ed.), *Science as Practice and Culture*. Chicago: University of Chicago Press.

———, and Yearley, S. 1992b. "Journey into Space." Pp. 367–389 in Pickering, A. (ed.), *Science as Practice and Culture*. Chicago: University of Chicago Press.

Cordes, C. 1994. "2 Scholars Examine the 'Bizarre War' against Science They Say Is Being Waged by the Academic Left." *The Chronicle of Higher Education*, April 27.

Davis, B. D. 1983. "Neo-Lysenkoism, IQ and the Press." *The Public Interest* (Fall):41–59.

Davis, B. D. 1986. *Storm over Biology: Essays on Science, Sentiment, and Public Policy*. Buffalo, NY: Prometheus Books.

Edge, D. 1996. Editorial. *Social Studies of Science* 26:723–732.

Feyerabend, P. 1992. "Atoms and Consciousness." *Common Knowledge* 1 (1):28–32.

Fraser, S. (ed.). 1995. *The Bell Curve Wars*. New York: Basic Books.

Fuller, S. 1995. "Two Cultures II: Science Studies Goes Public." *EASST Newsletter*, Spring.

Gitlin, T. 1995. *The Twilight of Common Dreams: Why America Is Wracked by Culture Wars*. New York: Metropolitan.

Gould, S. J. 1981. *The Mismeasure of Man*. New York: W. W. Norton.

Gross, P. 1997. "Opinion: The So-Called Science Wars and Sociological Gravitas." *The Scientist*, April 28:8.

———, and Levitt, N. 1994. *Higher Superstition: The Academic Left and Its Quarrels with Science*. Baltimore, MD: Johns Hopkins.

———, and Levitt, N. 1995. "Knocking Science for Fun and Profit." *The Skeptical Inquirer* 19 (2) March/April:38–42.

———, N. Levitt, and Lewis, M. (eds.). 1995. "The Flight from Science and Reason." *Annals of the New York Academy of Sciences* (775), 24 June.

———, N. Levitt, and Lewis, M. (eds.). 1996. *The Flight from Science and Reason*. Baltimore, MD: Johns Hopkins University Press.

Harding, S. 1991. *Whose Science? Whose Knowledge?* Ithaca, NY: Cornell University Press.

Herrnstein, R., and Murray, C. 1994. *The Bell Curve: Intelligence and Class Structure in American Life*. New York: Free Press.

Holden, C. 1996. "Anti-Evolution Show Prompts Furor." *Science* 271 (8 March):1357.

Hollinger, D. 1995. "Science as a Weapon in Kulturkampfe in the United States during and after World War II." *Isis* 86 (September):440–454.

Holton, G. 1993. *Science and Anti-Science*. Cambridge, MA: Harvard University Press.

———. 1994. "The Antiscience Problem." *The Skeptical Inquirer* 18 (3) Spring 1994:264–265.

Horgan, J. 1992. "Profile: Karl R. Popper; the Intellectual Warrior." *Scientific American* 267 (November):38–40.

———. 1993. "Profile: Paul Karl Feyerabend; the Worst Enemy of Science." *Scientific American* 268 (May):36–37.

Kevles, D. 1998. *The Baltimore Case: A Trial of Politics, Science, and Character*. New York: W. W. Norton.

Kohn, A. 1986. *False Prophets: Fraud and Error in Science and Medicine*. New York: Basil Blackwell.

Kuhn, T. 1970. *The Structure of Scientific Revolutions*. Chicago: University of Chicago Press.

Kurtz, P. 1994. "The Growth of Antiscience." *The Skeptical Inquirer* 18, 3 (Spring 1994):255–263.

Lakatos, I., and Musgrave, A. 1970. *Criticism and the Growth of Knowledge*. Cambridge: Cambridge Uiversity Press.

Latour, B. 1992. "One More Turn after the Social Turn." Pp. 272–294 in McMullin, E. (ed.), *The Social Dimensions of Science*. Notre Dame: University of Notre Dame Press.

Latour, B. 1993. *We Have Never Been Modern*. Harvard University Press.

Latour, B. 1995. "Who Speaks for Science?" (Letter). *The Sciences*, March/April:6.

Lawler, A. 1996. "Support for Science Stays Strong." *Science* 272 (31 May):1256.

Levins, R. and Lewontin, R. 1985. *The Dialectical Biologist*. Vambridge, MA: Harvard University Press.

Lewontin, R. C., Rose, S., and Kamin, L. 1984. *Not in Our Genes*. New York: Pantheon Books.

Longino, H. 1990. *Science as Social Knowledge*. Princeton: Princeton University Press.

Martin, B. 1996. "Social Construction of an 'Attack' on Science" (Review of *Higher Superstition*). *Social Studies of Science* 26:161–173.

Masterman, M. 1970. "The Nature of a Paradigm." Pp. 59–89 in Lakatos, I., and Musgrave, A. (eds.), *Criticism and the Growth of Knowledge*. Cambridge: Cambridge University Press.

McMillen, L. 1997. "The Science Wars Flare at the Institute for Advanced Study." *Chronicle of Higher Education*, May 16:A13.

Medawar, P. 1963. "Is the Scientific Paper a Fraud?" *The Listener*, 12 September:377–378.

———. 1969. *Induction and Intuition in Scientific Thought*. Philadelphia: American Philosophical Society; London: Methuen.

Numbers, R. 1992. *The Creationists*. New York: Knopf.

Popper, K. 1976. *Unended Quest: An Intellectual Autobiography*. Glasgow: William Collins Sons & Co.

Ruse, M. 1994. Review of Gross and Levitt's *Higher Superstition. The Sciences* 34 (November/December):39–44.

Schmaus, W., Segerstråle, U., and Jesseph, D. 1992. "The Hard Program in the Sociology of Scientific Knowledge: A Manifesto." Social Epistemology 6 (3):243–265. (With peer commentaries).

Searle, J. 1990. "The Storm over the University." *New York Review of Books*, December 6.

Segerstråle, U. 1993. "The Importance of Being Right vs. the Importance of Being Earnest: Public Accountability in the Baltimore Case." *Science Studies* 6 (2):4–22.

Segerstråle, U. 1995. "Good to the Last Drop? Millikan Stories as 'Canned' Pedagogy." *Science and Engineering Ethics* 1 (3):197–214.

Segerstråle, U. 2000. *Defenders of the Truth: The Battle for Science in the Sociobiology Debate and Beyond*. Oxford and New York: Oxford University Press.

Shils, E. 1986. Preface to B. Davis, *Storm over Biology: Essays on Science, Sentiment, and Public Policy*. Buffalo, NY: Prometheus Books.

Snow, C. P. 1959. *The Two Cultures*. Cambridge: Cambridge University Press.

Sokal, A., and Bricmont, J. 1997. *Impostures Intellectuelles*. Paris: Odile Jacob.

Steiner, G. 1971. *In Bluebeard's Castle*. New Haven: Yale University Press.

Stove, D. 1982. *Popper and After: Four Modern Irrationalists*. Oxford: Pergamon.

Theocharis T., and Psimopoulos, M. 1987. "Where Science Has Gone Wrong." *Nature* 329 (15 October):595–598.

Toulmin, S. 1990. *Cosmopolis: The Hidden Agenda of Modernity*. Chicago: University of Chicago Press.

Turney, J. 1993. "Science in the Real World." Review of *The Golem: What Everyone Should Know about Science. New Scientist*, 29 May:41–42.

Watson, J. 1968. *The Double Helix: A Personal Account of the Discovery of the Structure of DNA*. New York: Athenaeum Publishers.

Wilson, E. O. 1975. *Sociobiology: The New Synthesis*. Cambridge, MA: Harvard University Press.

Wilson, E. O. 1995. "Science and Ideology." *Academic Questions* 8 (3):73–81.

Woolley, M. 1995. "From Rhetoric to Reality." Editorial. *Science* 269 (15 September):145.

Zalewski, D. 1995. "Mad Scientists." *Lingua Franca* 5 (6) September/October:5–6.

Ziman, J. 1994. *Prometheus Bound: Science in a Dynamic Steady State*. Cambridge: Cambridge University Press.

CHAPTER 5

Anti-Antiscience: A Phenomenon in Search of an Explanation

Part II. The Conflict about the Social Role of Science

ULLICA SEGERSTRÅLE

EDITOR'S INTRODUCTION

In the United States, leading warriors against antiscience have gone further than merely warned about epistemological claims that could serve to undermine the social standing of science. The proscience activists have presented the various current criticisms of science as an attack on Reason itself—with vivid associations to earlier periods of appeal to irrational forces, notably the Nazi era.

It seems, then, that for these vocal scientists, science has taken on an extra cultural *dimension. In other words, just like the "postmodernists," they too argue that "science is culture," but they mean something quite different. It is, in fact, possible to see the Science Wars as a conflict about the cultural meaning of science. A postmodern-relativist-constructivist vision of science as "nothing but" culture or common sense is clashing with a vision of science as representing a* special, rational culture. *This, then, may be the real dividing line in the Science Wars, rather than a political left-right opposition. Many pro-science scientists in the Science Wars do in fact identify with the political* left. *This creates confusion, as happened for instance, in conjunction with the Alan Sokal hoax.*

One hidden problem in the Science Wars is both sides' seemingly unproblematic identification of 'reason' with science *or scientific rationality. What is missing is the recognition of a long-standing and complementary philosophical tradition of* practical reason, *that is, reason and rationality in human affairs.*

THE STRUGGLE ABOUT THE CULTURAL MEANING OF SCIENCE

In Part I, in pursuit of the meaning of "antiscience," I compared and contrasted recent episodes where academics have accused each other of "antiscience," finding continuities and discontinuities, consistencies and inconsistencies. Of particular interest were two episodes which appeared much stronger in their antiscience allegations than the Science Wars: recent nature-nurture controversies where scientific opponents to sociobiology and IQ testing were labeled "antiscience" and regarded as leaders of an antiscience movement, and a case in the United Kingdom in the late 1980s, where two physicists "proved" that it was Popper's philosophy that had caused the budgetary problems of British science.

Theocharis and Psimopoulos may have spelled out a possible train of thought when it came to the connection between the financial situation of science and the spread of antiscientific sentiments, but such a reasoning is *not* what we have directly heard from the fighters against academic antiscience in the United States. If anything, leading proscience warriors have explicitly denied that their concern would be motivated by funding problems (Gross, 1997; Weinberg, 1996). Moreover, as we have seen, the American anti-antiscience warriors appeared exclusively concerned with antiscience sentiments in *academia*, rather than with more widespread creationist or New Age beliefs, or even the poor state of scientific literacy in the general public. Apparently they thought that a forceful assault on the propagators of postmodern and constructivist critiques of science was the best way to stop the spread of antiscience sentiments and thus restore the natural state of affairs, which was pro science and reason.

An obvious way of looking at the Science Wars is, indeed, to see the scientists as finally striking back against the criticism of science from humanists and social scientists that had gone on for quite some time already. This kind of analysis may additionally invoke the struggle between the Two Cultures: here we would have one of the Two Cultures (science) vigorously protesting against the abuse it suffered at the hands of the other (the humanities). And, indeed, Gross and Levitt and certainly Alan Sokal (discussed in chapters 1 and 9) helped refuel this kind of impression of an opposition between the sciences and the humanities.

So let us explore this supposed struggle between the Two Cultures. If so, we may have to look further back in history. It is true that, from a perspective of the last twenty-five years or so, the Science Wars may, indeed, seem as the scientists striking back against the humanities (and social sciences). However, the situation may alternatively be described as (a few representatives for) the scientists trying to *recapture* something that it had recently lost to the other culture: the initiative as a progressive *cultural force*. For the proscience activists, science represented a crucially important aspect of culture and embodied important social values. It was this aspect of science that they perceived as being threatened.

Recently the historian of science David Hollinger addressed the central cultural role of science in the United States during and after the

Second World War (Hollinger, 1995). Although this proscience period is well-known, what is less known is that it was in fact the the outcome of a larger social struggle, a veritable *Kulturkampf*. In this cultural battle, the internationally minded scientific community was opposed by two other forces: the Protestant establishment and the then fascist-leaning Catholics. In this larger struggle between cosmopolitans and particularists science was seen as the embodiment of democratic values and rationality.[1] And science won the cultural battle. After the war, it was the values of science (objectivity, disinterestedness, fact-orientation, etc.) that were upheld as the model for culture in general in American society. Indeed, science became so much the model that an émigré logical positivist like Hans Reichenbach could seriously proclaim that most moral questions were potentially empirical questions that could be resolved with the help of science (Reichenbach, 1951, quoted in Hollinger, 1995, p. 447).[2]

In 1945, the idea of science-as-a-model-for-culture was strongly established in American academia. New science studies programs were institutionalized and "scientific" approaches to society were the order of the day (Hollinger, 1995; see also Barber's chapter in this volume). In 1945, a famous Harvard report, which became the basis of American higher education for the next two and a half decades

> explictly identified "science" as the foundation of "the spiritual values" of a democratic humanism and declared that American democracy needed citizens with "the habit of forming objective, disinterested judgments based upon exact evidence." (Hollinger, 1995, p. 445)

But this scientific or scientistic cultural ethos was not to remain unchallenged for long; already in the 1960s it met its first opposition from the "counterculture" movement, inspired by Theodor Roszak (1969), Herbert Marcuse (1964), and others. Roszak particularly clearly spelled out the opposition to "technocracy," scientific experts, and the root of it all: "objective consciousness." Interestingly, however, it does not seem that the scientific community perceived this as more than a passing nuisance. And these criticisms certainly do not seem to have affected the social position of science—even against the background of criticism, the scientific growth curve continued.

Thus, in the postwar United States the culture of science was strong and the battle for universalism at least temporarily won. But what about those traditional carriers of culture, the humanists? Hollinger depicts the humanities as staying out of the larger wartime cultural struggle altogether and sticking to strictly academic affairs. Meanwhile, in their actions (such as the awarding of literary prizes) they were in fact supporting just those reactionary values that the scientific community with its universalistic orientation abhorred. It was just this kind of contrast between the culture of the humanities and the culture of science that C. P. Snow was depicting in his *The Two Cultures*, Hollinger tells us.

This book, published in 1959, struck quite a nerve, we learn further. Because in his book Snow was not only saying what most people believe he said, namely that there was an unfortunate gap between the two cultures and that humanists had better learn thermodynamics. According to Hollinger, what Snow really argued was that in these two cultures, it was the *scientists*, not the humanists, who were the carriers of the true humanistic spirit! Snow accused the academic humanists of nothing less than betraying their cultural mission:

> He attacked a highly particular literary culture, the literary culture of modernism. . . . The faults he found with this culture, moreover, were less cognitive than moral. Snow accused the literati of perpetuating and celebrating a mythology of blood and history that had politically reactionary consequences. The modernists were basically cryptofascists, Snow implied, while the scientific professions carried "in their bones" a humane and democratic orientation toward the future. . . . Scientists are the vehicles for the open-minded, liberal-democratic, egalitarian values the whole world needs. (Hollinger, 1995, p. 448)

This was quite a charge. Obviously, the humanists were not letting Snow or anybody else get away with such a pronouncement. Indeed, as Hollinger points out, they soon enough struck back, wielding the names of thinkers such as Kuhn and Foucault as their weapon. And now it was the humanists' turn to tell the scientists that, far from being a liberating force of rationality, science was in fact restricted by prevailing dogma, or represented an oppressive Foucauldian knowledge-power system! And if that was so, then it was after all the *humanists*, rather than the scientists, who were the true representatives of freedom and democracy (Hollinger, 1995). In the light of this, we might understand the current criticisms of science within the humanities as part of the *humanists' delayed counterattack* against the postwar cultural hegemony of the sciences. In turn, it would be just against this counterattack of the humanities that Gross and Levitt and other fighters against antiscience were *reacting*.[3]

There is, however, a third player in the Science Wars, and that is the fundamentalist Christian churches, particularly the creationist movement. It seems that these conservative Christians, too, have recently picked up Kuhn and Foucault. For them, people like Kuhn and Foucault come in handy, because they can be used as "proof" against the epistemological authority of science and for the cognitive legitimacy of the Bible (Hollinger, 1995, p. 453). (The creationist strategy has long been to exploit any existing criticism of traditional science, particularly in regard to evolutionary biology.)

But it should be noted that this type of misplacement of target was equally true also for the other side in the Science Wars. Why were, for instance, the cultural critics of science not addressing society's power structures directly, rather than focusing on science as a symbol of power? On the face of it, then, it would appear that both sides in the current *Kulturkampf* had singled out the wrong target.

"ANTISCIENCE" AS A HEURISTIC DEVICE

But there may be a simple explanation for this apparent misidentification of targets on both sides. One reason why science had become such a popular subject in fields such as literary criticism and studies of rhetoric may be simply that humanists needed something *new* to analyze. Having exhausted standard subjects in literature (and, conceivably, the secondary literature on these as well), there was a desperate need for new material of some kind. And here, brilliantly, the humanists' newfound analytical tools were extended to scientific texts, a potentially inexhaustible source of ever new objects to study. However, although the humanists in this way decided to piggyback on science, their strategy as such was not novel at all. Indeed, we may say that the postmodern humanists acted just in the same way as many scientists: they took exciting-sounding new theories and methods and applied these to new fields of inquiry. Thus, what to some have seemed like a serious antiscience movement may have been largely the outcome of a clever professional innovation on the part of career-conscious academics.[4] And just in the same way, the new paradigm of social constructivism can be seen as a clever innovation in STS. For about three decades it generated a stream of case studies, fame for its protagonists, and quite a publication industry.

So much for some possible practical considerations underlying lofty theorizing among postmodern humanists and social constructivists. What might, then, have been a comparable mundane reason for the current opponents of antiscience to attack what they call the academic or cultural left? Although the fighters against antiscience can be said to be loosely united around a traditionalist belief in the integrity of science, they also seemed to have had their own *practical* reasons for forming a united front against the cultural critics of science. For instance, by joining the National Association of Scholars, academics from widely different fields who felt unfairly attacked by critics could find moral support from sympathizing fellow academics. At the same time, they could define themselves as together fighting the common enemy of "antiscience."

As we have seen, the label "antiscience" does not seem to be a good conceptual category. It groups too many heterogenous phenomena, and the fighters against antiscience have themselves sometimes disagreed as to just who belongs to the enemy camp. But, then again, who says that "antiscience" needs to be a conceptual category? There may be no need to look for the true meaning of "antiscience," if we simply interpret "antiscience" as a constructed category—more specifically, a collective *heuristic device* for unifying scientists in a collective protest against criticism of science. Under the general cry of "antiscience!" otherwise unconnected academics can strategically unite. They can also draw on a common repertory of anti-antiscience arguments. We saw already how E. O. Wilson anachronistically reconstructed his sufferings in the sociobiology controversy as due to assaults by "postmodern" radicals. (He also mentioned during a talk in 1995 that he had learned from the sociobiology controversy the importance of having allies—in that controversy many of his colleagues abandoned him.) Undoubtedly, many other scientists, too, attacked by cultural constructivists or other colleagues, were happy to join the protective pan-academic brotherhood of the National Association of Scholars.

It is interesting to compare the explanations of Gross and Levitt with those of Steven Balch, the president of the National Association of Scholars during its anti-antiscience campaigns. We are offered two quite different rationales for the association's decision to attack "antiscience" in 1994 and 1995. The authors of *Higher Superstition* presented the academic criticism of science as a problem for higher learning as a whole. While admitting that they themselves saw "little *early* change in the teaching and learning of science" as a result of this criticism, they still judged it "worthwhile to analyze at some length the animus against science currently expressed by the academic left." Why was that? The reason given was that "[i]ts existence has to be read as *the manifestation of a certain intellectual debility afflicting the contemporary university:* one that will ultimately threaten it." Thus, it was not only science that was under threat, it was the university itself! According to the authors, the health of the universities "has become incalculably important for the future of our descendants and, indeed, our species" (Gross & Levitt, 1994, p. 7, italics added). Presumably, then, it was for strictly intellectual reasons that Gross and Levitt (and Holton) were concerned with science and its critics and actively involved with the National Association of Scholars' conferences in 1994 and 1995.

However, the president of the National Association of Scholars himself expressed a less lofty and more practical view of the reason his organization had recently turned to science. In 1994 Balch told a reporter that the move to science came about in part because an important part (15%) of the association's 3,300 members were in science or medicine, and scientists had proven to be useful allies in fights over curricular changes in some universities (Heller, 1994). Thus, according to the president himself, there was a need to court especially scientists because of their academic standing and good fighting spirit!

Incidentally, the National Association of Scholars' interest in science may have been a rather temporary one. While, indeed, in 1994 and 1995 two conferences were held on the threat to science and reason, the next year the topic for the association's annual meeting was "The New Higher Education Reform Movement: Its Shape, Direction, and Future" (Leatherman, 1996). That was, indeed, a topic closer to the association's original agenda. Fighting antiscience was by no means the central goal of this organization: its real mission was defending objective standards and a traditional academic curriculum focused on "Western" values from threats of multiculturalism, "political correctness" and the like. The critique of natural science was thus a relative newcomer to this list.[5]

THE POLITICAL MEANING OF SCIENTIFIC OBJECTIVITY

Even though I am skeptical of a category called "antiscience," question the existence of an "antiscience movement," and point to heuristic reasons for mobilization of scientists, I do not question the genuineness of the concerns of the warriors against antiscience. The question is only the exact nature of these concerns. As we saw, intellectually and "logically," the hypothetical chain of thought of many current anti-antiscientists was identified already in the British discussion in 1987–1988. The reasoning was quite straightforward: it was the academic critique of scientific epistemology that caused the cut in science funding by undermining decision makers' trust in science. This may or may not have been true, but at least it was said, and could in principle be confirmed or refuted. Gross, Levitt, and others did not go this far, at least not in tangible statements, and it has been harder to grasp the exact nature of their quarrel with the cultural left.[6]

No doubt, one of the causes for Theocharis's and Psimopoulos's great distress was that in the 1980s the burden of proof had surreptitiously shifted in regard to the epistemology of science. These physicists perceived an apparent request for *scientists* themselves to do the unthinkable: convince a skeptical public of the epistemological merits of science and its ability to produce reliable knowledge! For them, all this had started when

Popper had incautiously pulled the plug out of the verificationist bulwark and thereby opened up the questioning of the very foundations of science. In contrast, the American warriors against antiscience seemed much less bothered by epistemological questions than Theocharis and Psimopoulos; they simply declared that "nothing is wrong with science." Indeed, what the present opponents to antiscience appeared to be really fighting for was something as esoteric-sounding as *the cultural authority of science*—or rather, a restoration of scientific values to their post–World War II social status.

It may not be necessary, after all, to ascribe further motives to the anti-antiscientists than exactly what they said themselves, namely that they feared "the flight from science and reason." Why all this talk about science and reason? For the new cultural critics of science, typically concerned with undermining science's claim to objectivity, it was probably not as obvious as for the anti-antiscience fighters that the claim to objectivity could serve as a formidable political weapon. Indeed, in certain historical cases, upholding *objectivity* was the *progressive* thing to do, particularly in a climate of nationalistic, conservative academic tradition. As seen above from Hollinger's account, after the Second World War, the cultural role given science entailed just this idea of objectivity as a political force against ideology.

In this sense, the Science Wars can be described as a conflict between two quite different total images of science. These were in turn connected to two different views of science's role in society and assessments of the present political situation. For the combattants of antiscience science was more than merely a rational and efficient way of obtaining knowledge about the world—for them, science had additional meaning as a *secular belief system* (cf. Holton, 1993, chapters 1 and 6). For this group of academics, scientific objectivity was tied to strong emotional values. But just in a similar way, science had emotional meaning, too, for many cultural critics of science, because they had identified science and objectivity with social oppression and elitism.

An illustrative example of the deep resistance felt by cultural left-oriented academics toward the very idea of objective science is the following recontruction of current "postmodern" anthropological reasoning, provided by Napoleon Chagnon, a subject of vicious attacks for his studies of the Yanomamö indians. He explains why it was that objectivist scientists were now seen as the real enemy by some anthropologists:

> The logic I have heard goes something like this: Most people now being studied by anthropologists are victims of oppression and live in bad conditions. Fieldwork now means that you are

> present at a crime scene and should become a witness whose first duty is to report crimes. You should use "interpretation" to identify crimes, not empirical data. This oppression is, in turn, caused by states. States wield power. Power rests on authority. Science is a kind of authority that the powerful will appropriate for oppressive goals. Therefore, scientific anthropologists are oppressors and are to be condemned for ethical, moral and even criminal wrongdoing. (Chagnon, 1995)

One indication that many members of the National Association of Scholars did indeed feel strongly about the value of scientific objectivity is the many comparisons between current science criticisms and values in Germany's Third Reich that were made during its science conferences. Holton had reportedly warned the Boston audience that "[w]e are now in a period in which there is a movement—not a well-synchronized movement—of delegitimizing science" (Heller, 1994). David Guston, commenting on the 1995 conference, "The Flight from Science and Reason" noted

> the near-incessant analogizing between constructivists and their irrational allies on one hand, and all sorts of evildoers, mostly of the highest order, on the other. A mere eleven minutes into the meeting, Paul Gross launched the first Nazi metaphor, linking the belief that "science is social" to Hitler's belief that "there is not such thing as truth." Although Gross immediately qualified his metaphor by denying any intention to assign membership in the National Socialist Party to social constructivists, the other speakers who rendered such analogies were not so polite. . . . Among the persons or belief systems explicitly linked by speakers to constructivism were: Hitler and Nazism; Stalinism and state Communism; Marxism and neo-Marxism; and Pontius Pilate. (Guston, 1995)[7]

What seemed to unite the fighters against antiscience, then, was a *strongly emotional belief in the political importance of objective science.* This was in principle true also for bona fide leftists among the scientists opposing antiscience. And here we come to an interesting paradox. In the continuous redefinition of the left-liberal position in the United States over the years, the "correct" left-liberal attitude to science had shifted as well. While at an earlier time science and arguments based on objective facts were seen as obvious political weapons for the *common* left-wing cause, this had more recently given way to a focus on the different political interests of different social subgroups instead. And here science *criticism* rather than objective science was one of the possible new political weapons. The reason the members of the National Association of Scholars were seen as conservatives by other academics may well have been that the idea of scientific objectivity had become antithetical to the new liberal credo in academia.

What was, then, Gross and Levitt's own political stance? At a panel discussion in conjunction with the 4S meeting 1995 in Charlottesville, Virginia, it became very clear that although Gross and Levitt were highly critical of such things as feminist theory of science (on this panel, they were debating Donna Haraway), they were generally *angry* with the current academic left. They seemed to feel that the left-wing forces in academia were being frittered away on unimportant bickering instead of concentrating on important issues. Levitt spelled out some of the issues. He wanted the left to study serious things that directly affected society: large-scale electric networks, regional planning, urban development, and so on, and the connections of these to money and power. After this, it was less hard to believe an early article saying that Gross and Levitt were in fact self-described *leftists* (Heller, 1994).

Thus, the Science Wars could at least partly be described as a clash between the "traditional" and the "cultural" academic left. However, what badly confused the issue was Gross and Levitt's insistence on calling the target of their critique "the academic left."[8]

Although Gross and Levitt in their book were rather cautious—perhaps not to alienate more conservative academics and thus build the broadest possible coalition—there were passages where their irritation with the present brand of cultural leftism came out quite clearly, as did their own wish for a return to serious left-wing concerns. They wrote, for instance, that "the left's flirtation with irrationalism, its reactionary rejection of the scientific worldview, is deplorable and contradicts its own deepest traditions" (p. 27). And toward the end of the book, they traced the "back-door utopianism" of the current academic left to a reactionary Romantic discontent, and saw "a tradition of egalitarianism falling under the sway of obscurantism and muddle" (p. 250). They noted that "*the left itself*—not only the peculiar ideological tribe we have dubbed the 'academic left,' but the far broader tradition of egalitarian social criticism that properly deserves such a designation—*is, potentially, one of the ironic victims of the doctrinaire science-criticism that has emerged . . .*" (p. 252, italics added). And, finally, they warned that "Scientists, and the scientifically well informed, will simply not accept any form of 'socialism' whose agenda include the subversion of legitimate science" (p. 252).

However, what made Gross and Levitt and other fighters against antiscience seem politically conservative was the latters' stubborn refusal to abandon their traditionalist belief that scientific epistemology is special and not on a par with other forms of knowledge. Furthermore, although the science warriors were fiercely interested in democracy, they did not share the cultural left's belief that *science* could be "democratized." Levitt and Gross in fact said as much in an article called "The Perils of Democratizing Science":

> Some bright people, competent in their own areas, really believe that because they *care* about AIDS or the environment or energy policy their concern will substitute for substantive knowledge. Enthusiasts and their lawyers wade into disputes about electromagnetic fields, genetic modification of food crops, the toxicity of breast implants, with little idea of methodological constraints, of plausible inference, or of the difference between a hypothesis and a conclusion. Nor are the courts always interested in those diffrences: Their primary job is to decide who wins and who loses.
>
> Such people do not mean to be antiscience. Their error is to convert the understandable feeling that science ought to be democratic to the fancy that democracy is a product simply of "getting involved." (Levitt & Gross, 1994)

And they went on to say the unutterable: that *science is not a democracy, and can never be.* According to them, "The reliability and utility of science depends on content, not on politics." Moreover, their view science is necessarily elitist: "[F]or science to dispense with the meritocracy of mind and skills that has served it so well would be to subvert itself and render it, ironically, less useful to democracy." So much, then, for the hopes of uniting democracy and science at the participatory level.

Obviously, then, the two camps had largely different conceptions about the *utility* of science. For the proscience camp, on the one hand, science served concrete social goals. It produced a reliable body of knowledge and the use of this knowledge was later negotiated within the democratic social process. For the cultural left, on the other hand, the primary interest was social power relations—including power relations within science—and, for it, *criticism* of science could be a useful tool in the struggle against political oppression of various minorities. What was at stake, then, in the discussion, may often not have been science at all, but rather, different conceptions of democracy. (In this respect, there was an interesting continuity with the 1960's left, explored in Segerstråle, 2000, chapter 17.)

Arguably, the cultural left did not really need this involvement with science: science criticism was just one of many possible tools to use; politics by other means. It was rather the more "traditionalist" members of the academic community for whom science was such a serious political matter. Because science was seen as so intimately connected to democracy, objective truth had to be upheld against "standpoint" epistemologies of various kinds. In this situation, however, it was in neither camp's interest to ask closer questions about the actual nature of science and scientific reasoning. In the first case, the black box of scientific knowledge was opened so wide that all the content flew away; in the second case, the box was snapped firmly shut in the face of anyone daring to ask questions. The mystery still remained how science actually works, a question that awaits its answer beyond the Science Wars.

THE MISSING REASON IN THE SCIENCE WARS

Several commentators noted the "irrationality" exhibited by the anti-antiscience warriors in their purported defense of reason (e.g., Guston, 1995; Levine, 1996). Others complained that Gross and Levitt did not uphold scholarly standards or failed to do their homework carefully, as would be expected of serious scientific critics (e.g., Edge, 1996). But Gross and Levitt did not claim that they acted as scientists when they criticized their academic colleagues in a popularly written book—they state explicitly that the aim with *Higher Superstition* is *polemical* (Gross & Levitt, 1994, p. 13). However, there was a more serious issue, which had to do with the very notion of "reason" employed by the anti-antiscience warriors—and their targets as well. Both sides in the Science Wars seemed to take for granted that the rationality of science was the only type of rationality that existed (this may have been connected to their emphasis on the mastery of natural science).

These perceptions may also have affected each side's efforts to explain the other side's motives. Gross and Levitt presented themselves as engaged in an "act of fairness" to understand the academic left (p. 217). On the one hand, they saw the left as motivated by career considerations and "envy" of the natural sciences. On the other hand, they complained that it was a contrarian or rebellious "mood" (which they described as a carryover from the 1960s) that made its members want to attack the most unique pillar of Western civilization: science (pp. 220–223). (In a surprising aside, Gross and Levitt even volunteered that their targets may have absorbed some of their "silliness" from the general cultural atmosphere—after all, full with creationist and New Age beliefs, p. 225!)

It is interesting to see the different tack taken by "cultural left" representative Andrew Ross, the editor of the *Science Wars* (Ross, 1996, an edited version of the special *Social Text* devoted to a response to Gross and Levitt, but overshadowed by Sokal's hoax), in his attempt to explain the science warriors. In his Introduction, he explained their actions as motivated by larger structural forces. The reason for the science warriors' attack on antiscience was really the closedown of the Supercollider, he argued, which signaled not only the end of Cold War science but also the more general downsizing of the economy and of Big Science. Ross charged that the emphasis on objectivity of science was just a cover: science had all the time successfully been connected up with the military and the industry. If anything, Ross noted, the compromise of academic science and its purported value of openness was only getting worse with the involvement of academics with industrial interests and patenting secrets. But, continued Ross, Gross and Levitt were supporters of conservative political interests that were not interested in "throwing the moneylenders out of the temple." Instead, Gross and Levitt accused others—the cultural constructivists—for spreading technophobia and irrationality.

In his next move, Ross brought up Ulrich Beck's *The Risk Society* (1986, 1992) to use this as a weapon against the proscience warriors. According to Beck, modernity had reached a point—late modernity—where it was turning against its own earlier, industrial phase. "Late modernity" was characterized by self-reflexivity, the attempt to use science to undo some of the damage of an earlier stage, such as environmental damage. At the same time in the "risk society" depicted by Beck, people were suffering from a widespread sense of threats from the environment and resentment of science and technology. But, Beck pointed out, scientific rationality could go only so far. For instance, people's perception of risk was something quite different from a scientific risk assessment model and could not be substituted with that. He concluded that scientific and technical experts had had too much social power; time had come to hear what the people wanted.

The reader might have expected Ross to continue arguing, that in challenging the expert power of technoscience the cultural left was just representing the reflexivity of late modernity. He might even have presented the academic left's criticism of science as *more modern* than the position of the current science warriors! But this was not what he did. Instead, Ross took another part of Beck's argument to its extreme, concluding that scientific rationality was fundamentally rotten and needed replacement. It represented only one point of view, a local Western knowledge that needed complementing with other types of knowledge, such as other local and traditional knowledges. What was needed, according to Ross, was a "science for the people."

But what was the position of those purported conservatives, Gross and Levitt? Were they cheering on technoscientific experts and their power? Not at all. In *Higher Superstition*, they too symphatized with people who were affected by the results of science and technology (p. 249). They too questioned the abuse of technoscientific power and its connection to the military and industry (pp. 224–225), but they also pointed to life-saving scientific discoveries (p. 229). They, too, said they wished for the public to participate better in discussions about science—although they wondered about the ways to instigate enthusiasm about science education in American culture (p. 248). They wrote about the undesirability of racism, sexism, and various types of discrimination in society, and in science (p. 216, p. 225). They even worried about better public participation in science policy (p. 251). What they did *not* seem to say was that scientists should run society because they knew better or were more rational. They seemed to be discussing democratic decision-making at the same time they talked about Science and Reason.

Even more ironically, also Gross and Levitt discuss late modernity. Their candidate was George Steiner, rather than Ulrich Beck, but it seems he was saying much the same thing, and earlier. Already in 1971, Steiner pointed out a fundamental paradox. Because of the self-reflexivity of modernity—which he described as a uniquely Western cultural feature—people at a later stage of modernity may come to condemn their own past (Gross & Levitt, 1994, p. 218). This means that Gross and Levitt *explicitly recognized criticism of science and society as legitimate.* According to them, however, the cultural left, caught in "the mood of late modernity," did not express important problems in clear language but muddled them with theory. In its hands, the critical reflection had got disconnected and "slipped free of reason" (p. 216). (So, Gross and Levitt did see the cultural left as representing this spirit of self-reflection—but they botched the job!)

What was going on here—save deliberate misreading and misrepresentation on both sides? I suggest that the problem had to do with the very meaning of "rationality." Both parties to the Science Wars seem to have been laboring under the same mistaken assumption that "rationality" was necessarily identical with "scientific rationality." Thus we had on the one hand the intense defense of something called "science and reason" against the dark forces of creeping irrationality, on the other an attempt to denigrate science and scientific rationality as an oppressive social force. For the proscience activists, science was single-handedly upholding social reason and order, for the cultural left, science was corrupt, aligned with social powerholders, and suppressing democratic discussion.

But the problem was that both sides gave too much power to *scientific* rationality. (This probably reflects an absence of a broader discourse in regard to different legitimate meanings of "reason" in contemporary academia.) What was not articulated by either side in the Science Wars, although both seemed to be indirectly acknowledging it, was that scientific rationality is not the only type of rationality. A philosophical tradition going back to Aristotle and including thinkers such as Kant, Weber, and today Jürgen Habermas, distinguishes between *two basic types of reason*:

instrumental and practical reason; functional rationality and substantial rationality, "systems" rationality versus normative discourse.[9] For Habermas—that indefatigable upholder of Enlightenment values and a veteran of fights with European postmodernists—the main concern has in fact been exactly to keep the two realms of rationality *separate*, that is, to keep the discourse about social goals and values from being "dominated" by the instrumental rationality characteristic of science and technology (Habermas, 1970, 1984, 1987).[10] Here Ross as well as Gross and Levitt appeared to be trapped in a too restricted all-or-nothing view of rationality while they did recognize the need for greater public involvement in regard to science. What appears to have been missing, then, in the Science Wars, is some recognition of this "rational discourse" about social goals and values (or what Habermas calls "communication ethics".)

The first thing to realize, then, is that 'rationality' can mean both scientific-technical discourse about means to an end, and a reasoned discourse about the very ends themselves. Or as, among others, Stephen Toulmin (1990) has pointed out, what scientific results we have is one thing, what we decide as a society to do with them is quite another matter.

That may be so in principle, but at least in the American setting, science is a such a strong social arbiter that scientific facts, even tentative ones, come to matter much more than they should. Science becomes the authority that can be invoked when social decisions are made. In the absence of a public discourse with well-articulated alternatives (and/or political parties to represent such opinions), the other type of social reason becomes invisible; it is not seen as existing in its own right. This is why we have the endless vigilance of potentially harmful effects of various types of scientific claims, such as the results of IQ research or research in biological aspects of human behavior—these are taken as scientific truths to be *acted* on. There is often the unfortunate combination of an exaggerated belief in the certainty of scientific claims with a keen orientation to practical application of them. Scientists, for various reasons, have not done much to dispell exaggerated trust in science (cf. Segerstråle, 1998, 2000, chapter 19).

Yet science is obviously needed as a socially important enterprise for organized truth-seeking and knowledge-production (cf. Fuchs's discussion in chapter 8). This may have been why Gross and Levitt did not want to tamper with science: "There is nothing wrong with it." What they did not explicitly emphasize was that science needed *complementing* with a different kind of social discourse, which might incorporate the various types of grievances of the cultural left while disconnecting these from a discussion about science. Gross and Levitt's vision of one type of rationality somehow embodied in science easily undermined their democratic ideas of public participation. They emphasized that the public needed to know science in order to have an informed opinion (p. 251), while they admitted that the American public was traditionally not interested in science (p. 248). This kind of position can be contrasted with, for instance, the one of Ulrich Beck, who believed that the public's concerns and fears should be taken seriously, *independently* of their knowledge of science.

Meanwhile, as the Science Wars raged, there continued to be great concern about the inherent *social authority of science*. (Gross and Levitt briefly touch on the power of expert knowledge on p. 248.) For instance, one recent position holds that, independently of how the public views science—as a progressive or a destructive force—scientists will still have cognitive authority in our culture. "And if there is anything that we should fear, it is the unexamined exercise of any form of authority, cognitive or otherwise" (Restivo & Bauschpies, 1996, p. 251; italics omitted).[11] Others had already earlier emphasized that science, just as any institution in a democracy, should not be outside democratic control:

> Without succumbing to anti-intellectualism, a democratic society must always be suspicious of knowledge in which the most valued forms of knowledge are the least accessible, or more sociologically, the most esteemed knowledge producers are the ones whose goods are accessible only to an elite set of consumers (e.g., other professional knowledge producers and, indirectly, their patrons; cf. R. Collins, 1979). (Fuller, 1992, p. 397)

None of this, however, seemed to mean that science needed to radically change; rather, it needed *complementing* with a good public discourse. Also, what was called for was a reassessment of its relationship to society and social needs (see, e.g., Byerly & Pielke, 1995, on "the changing ecology of science").

The Science Wars did not solve the problem of how to integrate scientific rationality with public discourse. However, one suggestion along these lines seems currently to exist in the field of risk analysis. The ambition there is to try to combine objective and subjective risk assessment models (Rosa, 1998), and to discuss public values in conjunction with existing technical alternatives.[12] Within a framework of "postnormal science"—a concept developed by Funtowicz and Ravetz (e.g., 1992, 1993), it is now argued that the public has to be brought in already at an early stage in discussions about science and society. In many cases it is no longer possible to leave the scientific and technical details to the experts for them to come up with ready solutions that can thereafter be democratically discussed. Instead, public perceptions will need to be consulted *during the very process* when decisions are taken about how to proceed.

In the Science Wars in general, there was altogether too much focus on "science" (as an idea, or an embodiment of Reason), while there was far too little real discussion about *science*, that is, the nature of science, or the relationship between science and social values. Let me end on an ironical note, invoking yet another notion of "science." Habermas has long upheld *scientific communication* as an exemplar of rational discourse; according to him, it has the required features and can therefore be used as a model for the "ideal speech situation," his ideal of rational social and political discourse (Habermas, 1970, 1994). This may be so (in the long run?), but in the adversarial climate of academic debate, this fact is certainly well hidden.

NOTES

1. For instance the sociologist Robert K. Merton figured prominently as a carrier of this message. (Merton, 1938/1973, 1942/1973). See also Bernard Barber's 1952 book *Science and the Social Order.*

2. Although Gross and Levitt may seem to be saying something similar with their mentioning "the public discourse of science and technology, a discourse already inadequate to the complexity of global issues at whose heart lie scientific questions " (Gross & Levitt, 1995), they in fact would like for the public to participate (see, e.g., Gross & Levitt, 1994, pp. 248–252).

3. This could explain some of the resentment of science within the cultural left that Gross and Levitt say they and other scientists have perceived (Gross & Levitt, 1994, p. 27).

4. According to Gross and Levitt, "[a] new and fashionable cottage industry has appeared among the intelligentsia, especially among academics. Its principal activity is to issue quantities of arrogant and hostile criticism of science" (Gross & Levitt, 1995).

5. It seems that the members of the National Association of Scholars so far form a small minority within academia, with little real impact on university campuses. A complaint at a recent conference was that they are often dismissed as reactionaries by other academics (Leatherman, 1996). However, the very fact that the National Association of Scholars exists indicates the deeper split in current American academia between those who radically want to change the university curriculum and those who support a traditional content. This is the gist of the so-called Culture Wars on American college campuses (see, e.g., Searle, 1990; Berman, 1992; Gitlin, 1995).

Incidentally, this split in regard to the criteria for "knowledge" has happened in many professional academic societies, too. For instance, a group called the Association of Literary Scholars and Critics reently seceded from the leading force among humanists, the Modern Language Association (MLA), and there have been similar trends in the organizations of anthropologists and historians (Leatherman, 1996; Heller, 1994).

6. In 1997, it seemed that the actual target of Gross was social studies of science, rather than postmodern humanists. In an article, he directly warned scientists against the dangers of STS (science and technology studies), suggesting that these people wish to control science and that they belong to a well-funded international movement (Gross, 1997). This apparent sharpening of the claws coincided with an academic spring campaign against historian of science Norton Wise's candidacy for a Research Professorship in Science Studies at the Princeton Institute for Advanced Studies. Although the details need to be sorted out, the blocking of Wise's position seems to have been part of an opposition by academic traditionalists against the field of science studies as such, going back to a time before Gross and Levitt's book. Some years earlier, Bruno Latour's appointment had been similarly blocked (cf. MacMillen, 1997).

7. In the edited proceedings of this conference (Gross, Levitt, & Lewis, 1996), there is little trace of the kind of atmosphere here described. In these proceedings, introductions and spoken comments were not included.

8. I hesitate to call these sections the "old" and the "new" left, because these terms already have their established meanings. The old left is understood to mean the earlier procommunist left in the United States, the new left came with the student upheavals in the 1960s. Generation-wise, Gross and Levitt are probably closer to the old left than the new left. Indeed, many scientists have felt a commonality between the universalist claims of science and communism; this was perhaps typically true for 1930s Britain, for people such as J. D. Bernal, the crystallographer and author of *The Social Function of Science*, or J. B. S. Haldane, the evolutionary biologist. (See Fuller's chapter for a discussion of the British scene.) With Sokal's hoax (Sokal, 1996a), and Sokal spelling out how his own left-wing position differed from that of the cultural left (Sokal, 1996b), matters became somewhat clearer (see the Introduction to this book).

9. Habermas grounds this second type of rationality, "communicative reason," on certain rules and expectations embedded in everyday language communication. (He is backed up by the particular branch of linguistics called "universal pragmatics," according to which our utterances can be seen as typically oriented toward potential normative validation by others: they imply descriptions, promises, threats, etc.)

Habermas wants to emphasize that we are born with the tools for attaining consensual truth through rational discourse: "What raises us out of nature is the only thing we can know: *language*. Through its structure, autonomy and responsibility are posited for us. Our first sentence expresses unequivocally the intention of universal and unconstrained consensus" (Habermas, 1971, p. 314), and "[T]ruth is built into the linguistic mechanisms of the reproduction of the species" (Habermas, 1984, p. 398). Under the unconstrained conditions of "the ideal speech situation," truth claims can be challenged and validated, and reasons can be requested and given in defense of any particular claim. In the end, what wins can only be "the better argument." In this way, at least in principle, rational discourse achieves truth through consensus.

10. In a famous essay, Habermas speaks about "Science and Technology as Ideology" (Habermas, 1968, 1970). He is there arguing against those who see science and technology as inherently evil. Instead, he suggests that science and the world of values occupy different complementary realms, which he calls "system" and "lifeworld" and which correspond to the two basic human interests in, respectively, environmental control and human understanding. The types of rationality prevailing in these realms are therefore also differently grounded. Science and technology become "ideology" exactly when their type of rationality is seen as the *only* valid approach:

> Technocratic consciousness reflects . . . the repression of "ethics" as such as a category of life. The common positivist way of thinking render inert the frame of reference of interaction in ordinary language . . . The reified models of the sciences migrate into the sociocultural lifeworld and gain objective power over the latter's self-understanding. The ideological nucleus of this consciousness is *the elimination of the distinction between the practical and the technical.* (Habermas, 1970, p. 112–113)

11. See Restivo and Bauschpies (1996) for a collection of intellectuals who have pointed to dangers of making science the supreme cognitive authority.

12. An example of this is, for instance, the Canadian authorities' recent detailed field study of public opinion about reproductive technologies, whose aim was to inform the orientation of future research and legislation in this field (Lori Andrews, spoken Introduction to "Changing Conceptions: A Symposium on reproductive Technologies," Institute for Science, Law and Technology, Illinois Institute of Technology, December 5, 1997).

REFERENCES

Beck, U. (1992/1986). *Risk Society: Toward a New Modernity.* London: Sage Publ.

———. (1995/1991). *Ecological Enlightenment: Essays on the Politics of the Risk Society.* Atlantic Highlands, N.J.: Humanities Press.

Berman, P. (ed.) (1992). *Debating P.C.* New York: Bantam Doubleday Dell.

Bloor, D. (1997). Remember the Strong Program? *Science, Technology and Human Values* 22 (3): 373–385.

Byerly, R., and Pielke, R. 1995. "The Changing Ecology of United States' Science." *Science* 269 (15 September):1531–1532.

Chagnon, N. 1995. "The Academic Left and Threats to Scientific Anthropology." *Human Behavior and Evolution Society Newsletter* 4 (1): 1–2.

Collins, R. 1979. *The Credential Society.* New York: Academic Press.

Edge, D. 1996. Editorial. *Social Studies of Science* 26:723–32.

Fuller, S. 1992. "Social Epistemology and the Research Agenda of Science Studies." Pp. 390–428 in Pickering, A. (ed.), *Science as Practice and Culture.* Chicago: University of Chicago Press.

Funtowicz, S., and Ravetz, J. 1992. "The Good, the True, and the Post-Modern." *Futures* 24: 963–976.

Funtowicz, S., and Ravetz, J. 1993. "Science for a Post-Normal Age." *Futures* 25:739–752.

Gitlin, T. 1995. *The Twilight of Common Dreams: Why America Is Wracked by Culture Wars*. New York: Metropolitan.

Gross, P. 1997. "Opinion: The So-Called Science Wars and Sociological Gravitas." *The Scientist*, April 28:8.

Gross, P., and Levitt, N. 1994. *Higher Superstition: The Academic Left and Its Quarrels with Science*. Baltimore, MD: Johns Hopkins.

Gross, P., and Levitt, N. 1995. "Knocking Science for Fun and Profit." *The Skeptical Inquirer* 19 (2) March/April:38–42.

Gross, P., N. Levitt, and Lewis, M. (eds.). 1996. *The Flight from Science and Reason*. Baltimore, MD: Johns Hopkins University Press.

Guston, D. 1995. "The Flight from Reasonableness." (Report from the "The Flight from Science and Reason" conference). *Technoscience* 8 (3), Fall:11–13.

Habermas, J. 1970. *Toward a Rational Society*. Boston: Beacon Press. (Originally publ. in German, Frankfurt: Suhrkamp, 1968.)

———. 1971. *Knowledge and Human Interests*. Boston, MA: Beacon Press. (Originally publ. in German, Frankfurt: Suhrkamp, 1968.)

———. 1984. *The Theory of Communicative Action. Volume I: Reason and the Rationalization of Society*. Boston, MA: Beacon Press. (Originally publ. in German, Frankfurt: Suhrkamp, 1981.)

———. 1987. *The Theory of Communicative Action. Volume II: Lifeworld and System*. Boston, MA: Beacon Press. (Originally publ. in German, Frankfurt: Suhrkamp, 1981.)

Heller, S. 1994. "At Conference, Conservative Scholars Lash Out at Attempts to 'Delegitimize Science.' " *The Chronicle of Higher Education*, November 23: A18, A20.

Hollinger, D. 1995. "Science as a Weapon in Kulturkampfe in the United States during and after World War II." *Isis* 86 (September):440–454.

Holton, G. 1993. *Science and Anti-Science*. Cambridge, MA: Harvard University Press.

Latour, B. 1987. *Science in Action*. Cambridge, MA: Harvard University Press.

———. 1992. "One More Turn after the Social Turn." Pp. 277–294 in McMullin, E. (ed.), *The Social Dimensions of Science*. Notre Dame: University of Notre Dame Press.

———. 1993. *We Have Never Been Modern*. Harvard University Press.

Lawler, A. 1996. "Support for Science Stays Strong." *Science* 272 (May 31):1256.

Leatherman, C. 1996. "Conservative Scholars' Group Finds Itself with New Allies." *The Chronicle of Higher Education*, May 17:A22.

Levine, G. 1996. Letter. *The New York Review of Books*, October 3:54.

Levitt, N., and Gross, P. 1994. "The Perils of Democratizing Science." *The Chronicle of Higher Education*, October 5:B1–B2.

Marcuse, H. 1964. *One Dimensional Man*. Boston: Beacon Press.

Merton, R. K. 1938/1973. "Science and the Social Order." Pp. 254–66 in Merton, R. K., *The Sociology of Science*. Chicago: The University of Chicago Press. (Republication from *Philosophy of Science* 5 (1938):321–37.

Merton, R. K. 1942/1973. "The Normative Structure of Science." Pp. 267–278 in Merton, R. K., *The Sociology of Science*. Chicago: The University of Chicago Press. (Republication of original, entitled "Science and Technology in a Democratic Social Order," *Journal of Legal and Political Sociology* 1 (1942):115–26. Later published as "Science and Democratic Social Structure" in Merton, R. K., *Social Theory and Social Structure*, New York: The Free Press, 1949, 1957).

Reichenbach, H. 1951. *The Rise of Scientific Philosophy*. Berkeley: University of California Press.

Restivo, S., and Bauschpies, W. 1996. "Science, Social Theory, and Science Criticism." *Communication and Cognition* 29 (2):249–272.

Rosa, E. 1998. "Metatheoretical Foundations for Post-Normal Risk." *Journal of Risk Research* (forthcoming).

Ross, A. 1996. *The Science Wars*. Durham, N.C.: Duke University Press.

Roszak, T. 1969. *The Making of a Counter Culture: Reflections on the Technocratic Society and Its Youthful Opposition*. New York: Doubleday.

Searle, J. 1990. "The Storm Over the University." *New York Review of Books*, December 6.

Segerstråle, U. 1998. "Truth and Consequences in the Sociobiology Debate and Beyond." Pp. 249–281 in Falger, V., Meyer, P., and van der Dennen, J. (eds.), *Sociobiology and Politics, Research in Biopolitics*, Vol. 6. Stamford, CT, and London: JAI Press.

Segerstråle, U. 2000. *Defenders of the Truth: The Battle for Science in the Sociobiology Debate and Beyond*. Oxford and New York: Oxford University Press.

Shapin, S., and Schaffer, S. 1985. *Leviathan and the Airpump*. Princeton, NJ: Princeton University Press.

Sokal, A. 1996a. "Trangressing the Boundaries: Toward a Transformative Hermeneutics of Quantum Gravity." *Social Text* 46–47:217–252.

———. 1996b. "A Physicist Experiments with Cultural Studies" *Lingua Franca* 6 (4):62–64.

Toulmin, S. 1990. *Cosmopolis: The Hidden Agenda of Modernity*. Chicago: University of Chicago Press.

Weinberg, S. 1996. "Response." *The New York Review of Books* October 3:55–6.

CHAPTER 6

Visions of Science in the Twentieth Century

VALÉRY CHOLAKOV

EDITOR'S INTRODUCTION

In this chapter we move to examine the historically changing social support for science. Under what conditions do we typically have proscience or antiscience? Valery Cholakov, a historian with training in science, and the author of The Nobel Prizes, *examines three cases of strong proscience sentiments: post–World War II United States, the former Soviet Union, and pre–World War I Germany. These all also came with their own metaphors: "science, the endless frontier," "science, a direct productive force," and "science, the land of opportunity," respectively. These slogans, which presented science as a positive social force and problem solver, were tied not only to national security interests, but also to science's ability at that point in time to make dramatic and visible contributions to people's everyday life.*

CHANGING VIEWS OF SCIENCE

Ever since the beginning of the modern period almost five centuries ago we have contrasting attitudes toward science. On the one hand there is a fascination with knowledge and the power over nature. On the other hand, there is a rejection of science based either on a presupposed inability of science to find the truth, or, conversely, on a fear of science being too efficient. If we focus on the period after World War II, we can easily identify these alternative attitudes toward science. (For a discussion of earlier periods of proscience and antiscience, see, e.g., Holton, 1993, chapter 6.)

The optimism after World War II led Vannevar Bush to proclaim science as "the endless frontier," which would lead humanity to progress and prosperity (Bush, 1945).[1] About a decade later, in 1956 in the former

Soviet Union, science was declared "a direct productive force." (In fact, hardly any other country has put so much hope to science as the former U.S.S.R.) But at the same time the Cold War was gaining speed, involving many scientists in weapon making. This created a reaction against science: the critics feared that science, and the technology it produced, might slip out of human control and endanger the very existence of the human race (e.g., Kevles, 1987, chapter 16.)

Positive or negative, both attitudes were based on the same belief in the omnipotence of science. However, by the 1970s a new form of critique of science emerged, this time in the form of philosophical relativism and social constructivism. Unlike the earlier critics, the new critics challenged the very foundations of science: its ability to find truth and solve problems. This critique gained momentum and reached a peak in the postmodernist era of the 1980s. To many scientists and to philosophers and sociologists trained a generation earlier, this seemed as a wave of obscurantism and a real antiscience movement. In the light of these recent developments, it is interesting to ask the question: What produces certain attitudes to science, either for or against?

SCIENCE: THE ENDLESS FRONTIER

We can get some guidance by examining cases of proscience or antiscience sentiments and movements in their historical context. For instance, the famous slogan of Vannevar Bush, "science the endless frontier," can be linked to a period in the history of the United States, which historians of technology have characterized as a period of "technological optimism." Generally speaking, this is the time between 1870 and 1970, as depicted by the historian Thomas Parke Hughes in his book *American Genesis* (Hughes, 1989). However, while the spirit of this century in American history may be uniformly regarded as one of technological optimism, historically speaking it is rather complex. Periods of economic growth interchanged with stagnation, the country moved through modernization, urbanization, and demographic transition, and the nation participated in two World Wars. This is also reflected in the shifting attitudes to science during this time.

Vannevar Bush's famous manifesto was published in 1945, but it marks a process that began much earlier. The immediate precursor of the developments in the postwar period was the period between the two World Wars, and it is in fact this time that is the best example of high beliefs in science and its ability to solve social problems. While science could not avert the Great Depression, during that era there was a proliferation of projects for bringing the country to prosperity by the means of scientific discovery. Indeed, during the worst times of the Depression in 1933 and 1934, the famous exposition Century of Progress was organized in Chicago. There, in a most direct way, millions of people were introduced to technological marvels, which promised to change human existence (Rydell, 1993). All this bright future was to come about through the application of science to the economy and everyday life.

Scientific fairs have a long history, starting with the English Crystal Palace exhibit of 1851. In the 1930s, however, they acquired a special meaning throughout the industrialized world: they were to assure the million victims of the Great Depression that the bad times would be overcome by the magic of science. During the same time, the Soviet Union was experiencing a comparable social and economic crisis: the pains of rapid industrialization. There, too, science exhibits flourished. A testimony to the faith in science in the Soviet Union was that one scientific exposition in Moscow was made permanent and lasted until the end of the Soviet Union.

But it was not really science that rescued these countries from their economic problems. The real restructuring of the economy occurred in fact during World War II through the enormous efforts of the fighting nations. This war also saw the implementation of a major discovery made in a very abstract scientific field: atomic physics. Indeed, until the mid-30s atomic physics was considered too theoretical and remote from practical applications even by some famous scientists, like Rutherford.[2] In the Soviet Union, scientists were simply forbidden to work in atomic physics and ordered to go into more practical research instead. However, by the beginning of World War II, all of a sudden new discoveries made atomic physics a hot field, resulting in the atom bomb in 1945.

A good question to ask is whether it was this unexpected outcome from an abstract area of science that made Vannevar Bush insist that pure science should have preeminence over applied research. In any event, this was to become the paradigm of science policy throughout the next fifty years. It is only recently that the wisdom of Bush's pure science idea has come to be challenged. For instance, a recent article in *Science* called "The Changing Ecology of United States Science" appealed for a return to the interwar concept of "useful knowledge" instead of "pure science" as a guideline for science policy (Byerly & Pielke, 1995).

SCIENCE: A DIRECT PRODUCTIVE FORCE

The "useful knowledge" approach was something that flourished particularly in the former Soviet Union. It was exactly by promising useful knowledge that the old Imperial Academy of Sciences in St. Petersburg was able to successfully transform itself into the new Soviet Academy of

Sciences in Moscow. Through this maneuver, this remarkable scientific body managed to regain state patronage and retained considerable autonomy despite the zigzags of ideology (Vucinich, 1984). In the postwar period the concept of preeminence of basic research gained some currency in the rhetoric of the Soviet Union. Still, science continued being considered a "direct productive force." This became the official doctrine in 1956.

Enthusiastic belief in science had existed in other nations before, but hardly on such a scale as in the Soviet Union. There science was believed to be the key to economic success in a truly Baconian fashion. Popular science journals published in millions of copies every month provided a beautiful world picture, just as Tomaso Campanella might have wished it.[3] Indeed, the reliance on science was so strong that scholars from various fields even developed the special discipline of "science of science," which was supposed not only to study science, but also to steer it scientifically.

THE SCIENCE OF SCIENCE

The science of science movement started in 1959 in Prague, where the twentieth anniversary of John Desmond Bernal's book *The Social Function of Science*, was celebrated. While the slavic term for "science of science" actually originated in Poland in the 1930s and etymologically resembles the term "Wissenschaftslehre" coined by Bernard Bolzano in the 1830s, the Bernalian connection marks the new use of this concept. John Bernal was part of the group of leftist British scientists in the 1930s, whom Gary Werskey named "The Visible College" (Werskey, 1978). Indeed, many of them participated directly in politics and advocated a socially responsible science.

Bernal's book, which is as structured as a treatise on crystallography might be, ends with a paragraph "Science as Communism" (Bernal, 1964).[4] While it could be argued that there was some Soviet connection with this British development, the influence may actually have come from France. Indeed, the famous essay of Boris Hessen at the History of Science congress in England in 1931 made a connection between historical formations and the internal development of science (Hessen, 1931).[5] Visitors to the Soviet Union, like Julian Huxley in 1930, wrote sympathetic accounts of Soviet science and the prospects it opened for the economy (Huxley, 1930). According to Terry Shinn, however, it was the French leftist program for science buildup of the 1920s that influenced Bernal and the other British scientists. The French scientists wanted the establishment of a special institution for research amounting to nothing less than a state within the state, in the words of Shinn (Shinn, 1994). After 1936, when a left-wing government came to power in France, this project became viable, and indeed, Centre National de la Recherche Scientifique (CNRS) was established in 1939. The institution resumed work in 1945. This became the model of scientific organization for the Comecon countries to follow from the 1960s to the 1980s.

The term "science of science" had some currency as well in other countries, although in the West, it was mostly believed that prosperity through science would come about by more democratic means and that science needs only sponsoring and developing along Western academic traditions. A volume from 1964 on this topic (Goldsmith & Mackay, 1964) contains a cautious assessment of science of science by Derek Price (Price, 1964) along with chapters discussing how Third World countries would achieve modernization and prosperity if they only developed their national science.[6]

CHEMISTRY: THE LAND OF OPPORTUNITY

"Science the endless frontier," and "science as a direct productive force" were the two famous slogans from the post–World War II era. But the number of examples of technological optimism could be expanded. One such case is the development of chemurgy in the interwar period in the United States. The name translates as "action by the means of chemistry." It was a program for scientific and sociopolitical development aiming to rescue American agriculture from depression and to boost the economy into a new era of growth and expansion. This case is interesting because of the fact that it was not linked to the state bureaucracies or the military-industrial complex. Indeed, among the sponsors of chemurgy were self-made men and famous capitalists such as Henry Ford.

Chemurgy emerged as a reseach program in the late 1920s, when American agriculture began to suffer from overproduction. This brought about ideas to use farm products as raw material for the chemical industry. According to the promoters of these projects, scientific research and technological development could create new industries, which could absorb agrarian surplus and make the farmers prosperous. The prosperous farmers, then, would start spending on industrial goods, which in turn, would stimulate industry. Thus, in the end, economic growth and prosperity could be achieved by the implementation of science (Borth, 1939). This program enjoyed enormous popularity in the 1930s, when 25 percent of the American population were farmers and millions of people suffered unemployment because of the Depression. As such, also this program can be interpreted within the earlier mentioned concept of "useful knowledge."

In reality, however, chemurgy did not fulfill its promise—perhaps for reasons lying outside science. But there was another scientific program, which turned out to be quite efficient—German chemistry. Throughout the first third of this century, Germany developed technologies to synthesize materials needed by its economy in peace and war. This enabled the country to survive the era of imperialism and two world wars and gave rise to the German expression: "chemistry—the land of opportunity." The phrase was coined by Emil Fischer in 1911 at a meeting attended by the Kaiser in 1911 (Johnson, 1990). "The land of opportunity" was actually a paraphrase of the thesis about the "endless frontier," a vision of American history which gained popularity in the 1890s with Frederick Jackson Turner. Thus, the German concept in fact predates the slogan of Vannevar Bush by more than thirty years. It is not surprising that for Germany, which did not have a large territory like the United States or Russia or a colonial empire like Britain or France, science would be seen as the means to economic progress.

In Germany chemistry has for centuries been tied to national interests. This tradition begins with Glauber in the seventeenth century and continues with Liebig in the mid-nineteenth century, and with Haber, Bosch, Bergius, and others in the early twentieth century. Liebig appealed to German chemistry to be socially responsible—he called the chemistry of his days "asocial" for being too theoretically oriented and not focused on the needs of society. While Glauber and Liebig can be linked to the study of fertilizers and the improvement of the productivity of German agriculture, others developed synthetic chemistry, which made Germany almost self-sufficient in regard to raw materials.

Thus Germany, a latecomer in the industrial world, sought to compensate for the lack of colonies and natural resources by fostering chemical synthesis. The impact of some of these scientific developments was of a global nature. By the end of the nineteenth century, Adolf von Baeyer was able to synthesize indigo, which led to the collapse of the British indigo plantations in India (Bud, 1994). The synthetic dye industry had emerged with Perkin in England but was brought to a high art in Germany. This success, in turn, brought about attempts to solve other problems with raw materials of prime importance. The Haber-Bosch process of nitrate synthesis eliminated Germany's dependence on imports from Chile and provided ammunition for World War I. The most famous cases are the ammonia synthesis developed by Haber and Bosch, and the synthesis of gasoline from coal developed by Bergius. All of these scientists received Nobel prizes, thus testifying to Alfred Nobel's own vision of useful knowledge (Cholakov, 1986).

Remarkably, Bergius even managed to produce sugar from wood. (As a demonstration, he was handing out candy made of hydrolyzed wood to visitors to his laboratory; Robert Bud, personal communication). In 1939, Bergius's success prompted an American journalist to snappily remark that now, in a case of war, Germany could eat its forests! (Borth, 1939). Indeed, in the 1930s, Bergius's efforts were seen as a national achievement. A biography of Friedrich Bergius published in Berlin in 1943 bears the title: "Ein Deutscher Erfinder kämpft gegen die Englishe Blokade" (A German Inventor Fights against the English Blockade) (Schmidt-Pauli, 1943). It has two parts: I. Kraftstoff aus Kohl (Fuel from Coal), and II. Nahrung aus Holz (Food from Wood).

The late nineteenth century, labeled by its contemporaries as an age of imperialism, saw a divided Earth and a struggle for resources. In such restrained conditions, it is little wonder that different nations put their hopes in science and perceived it as alternatively the land of opportunity, the endless frontier or a direct productive force. The time between the mid-nineteenth and the mid-twentieth centuries saw rapid modernization and industrialization, first in Europe and then in other parts of the world. Change became part of the everyday experience for millions of people and the claim of scientists that it was science that had brought about the change found a ready audience among both masses and elites. Meanwhile, in some cases the scientific establishments did provide crucial support to their nations, for example in England and Germany during World War I and the United States during World War II.

SCIENCE: A DANGEROUS POWER

But just as, historically, technological optimism was produced by modernization, the pains of industrialization generated antiscience sentiments. A famous early example is Mary Shelley's *Frankenstein* from 1818; a later one is Jacques Ellul's writings about the power of technology over man in 1954, a book that was translated into English only in 1964 (Ellul, 1964). These antiscience views and fear of technology typically reflect a fear of change and a nostalgia for the rural past. Present-day environmentalism is one of the offsprings of these sentiments.

The beginning of the Great Depression in 1929 brought doubts about the benefits of science. By 1930 there were calls in Britain and the United States for a moratorium on further research in science and mechanization in industry (Mendelsohn, 1994). These fears would resurge again in the 1960s, when automation seemed imminent and are still perpetuated by writers like Edward Luttwak, who believes that the danger of automation has only been postponed (Luttwak, 1994).

In the period after World War II, there was another line in antiscientism, indirectly engendered by the Manhattan project. Big science came increasingly to be regarded as big power. Some of the social scientists in the 1970s and 1980s came to see themselves as fighters against powerful establishments entrenched in science and politics. By exposing the shortcomings of science, they believed that they could undermine the claims and legitimacy of those very establishments. This made sense during the times of the Cold War, when to many the world seemed to be on the brink of destruction.

This brings us to one of the uses of science that is not much talked about: the link between science and the military. Of course, this connection can be traced back to Archimedes, and later to Renaissance figures like Leonardo da Vinci and Galileo. In search of favors and patronage, scientists offered to princes not only gold-making but also better weaponry. The French state incorporated both science and engineering in the military, and so did the Russian empire. In liberal nineteenth century France, however, research in powder-making was left predominantly to individuals like Alfred Nobel, who was not even a national (Cholakov, 1986).

World War I once again changed the link between science and the state. During it science played a primary role, either through German chemistry (nitrate explosives), British microbiology (rubber made by the Weizmann method) or French medicine (preventing epidemics among soldiers). After the war, however, the link dissolved. Probably only in Russia did this effort to continue research for the aims of national security persist without interruption. World War II again mobilized scientists for state and military service. This time there was an unexpected discovery: atomic fission, which led to the creation of terrifying weapons and an arms race.

It is useful here to take one more look at *Science the Endless Frontier*. Vannevar Bush himself was one of the architects of the mobilization of American science for the military effort. When he, in his manifesto, reiterated the promises of science in solving the problems of society, he had also something else in mind than the abstract notion of useful knowledge. He wanted to preserve the scientific establishment that had emerged during the war.

In American history there is a pattern of postwar slumps and depressions, due to the disappearance of the war-time demand for production. This is also what happened in the mid-1940s. However, by the end of the decade, the Cold War and the Korean War mobilized the nation once again—indeed, as Dean Acheson famously put it, "The Korean War saved us" (Yergin, 1977). Such was the buildup throughout the 1950s that in 1959 President Eisenhower spoke with concern about the power of the military-industrial complex.

THE END OF BIG SCIENCE

After the war, however, Vannevar Bush had chosen to speak for the preeminence of pure research over applied research and argue for pure research as the road to useful knowledge. Indeed, for several decades, scientists built larger and larger accelerators, larger and larger telescopes, explored the oceans and sent men and probes into outer space. While this was perceived with great enthusiasm by the people of the United States, the Soviet Union and other nations, eventually a point of saturation was reached. With the end of the Cold War, many of these areas of research are shrinking. In 1989 the Berlin wall, the very symbol of the Cold War, was torn down. Some time after that, far away in Texas, the greatest project of postwar science, the Superconducting Supercollider (SSC) ran into trouble, and was eventually dismantled. To many scientists, the end of the SSC signified a changed attitude to science in society. Nobel prize winners like Leon Lederman spoke about "the end of the frontier" (Lederman, 1991).

However, this leaves those who witnessed the post–World War II developments with the question: Did the accelerators, in fact, model nuclear reactions in exploding bombs? Did the big telescopes watch for enemy satellites? Did the oceanographers scout for somebody else's submarines, or did the landing on the Moon symbolize the success of rocketry? These questions certainly are getting coverage in recent scholarship, and the Faustian bargain between pure science and military research may become more explicit (Leslie, 1993, 1994). In this connection the attack on science during the postmodernist 1980s may be perceived as an attack on *power*, quite in the spirit of the leftist and radical tradition. Indeed, in the time of Star Wars, science was so much identified with power that some academics may have believed that they could change the world through political activism within science—or by devoting themselves to science studies.

Meanwhile, if Big Science was indeed primarily linked to military projects and the end of the Cold War removed some of this type of science from the nation's top priority list, then one might argue that recent antiscience sentiments generated by social scientists and environmentalists may, in fact, have been skilfully exploited by politicians to cut the science budget. (In 1987, the physicists Theocharis and Psimopoulos made this type of argument for science in Britain; for more detail, see Segerstråle's chapter 4 in this book.)

Now the Cold War is over, the budgets for research are cut throughout the world, and we are all somewhat bored with science. Now what? The exponential growth of Big Science, as observed by Derek Price in the early 1960s (Price, 1963), has come to an end—a situation that John Ziman describes as "science in a steady state" (Ziman, 1994). But the case is not only that pure and applied research are more and more fused, as are also the boundaries between science departments and corporate structures. Today's scientist may turn out to be also a venture capitalist, a patent holder, a developer, and a salesman. Science has become a sort of high-tech craftsmanship.

REDEFINING SCIENCE

While this may seem as an end of an era to some, one could argue that this steady-state science is actually a return to the early modern condition and the Baconian program. Science is finally becoming a direct productive force. The artificial bubble of the Cold War era is gone, but that will only put science to solve more of the real problems of the world. If we draw a historical parallel, today's receding enthusiasm about science is similar to eighteenth-century England, where science turned into a technicality during the Industrial Revolution. In this respect the grand science of the recent past will look similar to the state-sponsored and militarized science of the ancien régime which failed to reform itself and fell victim to a revolution. We probably have little to regret about it and everything to hope for.

NOTES

1. Vannevar Bush, who became the director of the Office of Scientific Research and Development in 1941 wanted to sustain the new scienific establishment after World War II was over.

2. Ernest Rutherford (1871–1937) was instrumental in the discovery of atomic energy, but he was skeptical about the practicality of its use.

3. As Margaret Jacob points out (Jacob, 1988), Tomaso Campanella proposed to the King of Spain to distract people with stories about nature in order for them not to go into politics.

4. It is an interesting question to ask whether his work in crystallography affected his general thinking about science and society.

5. Hessen's essay is believed to have been very influential among English intellectuals at the time (e.g., Werskey, 1978).

6. While Price liked the idea of science of science, he was weary about turning it into management of science.

REFERENCES

Bernal, J. D. 1964. *The Social Function of Science*. Cambridge MA: MIT Press (With the essay "After 25 Years." First published in 1939).

Borth, C. 1939. *Pioneers of Plenty: The Story of Chemurgy.* Indianapolis, New York: Bobs Merrill Co.

Bud, R. 1994. *The Uses of Life. A History of Biotechnology.* Cambridge: Cambridge University Press.

Bush, V. 1945. *Science the Endless Frontier. A Report to the President.* Washington, D.C.: U.S. Government Printing Office.

Byerly, R., and Pielke, R. 1995. "The Changing Ecology of United States Science." *Science* 269 (15 September):1531–1532.

Cholakov, V. 1986. *Nobelevskie premii.* (The Nobel Prizes). Moscow.

Ellul, J. 1964. *The Technological Society.* New York: Alfred Knopf.

Hessen, B. 1931. *Science at the Crossroads.* London: KNIGA. (Reprinted as Science at the Crossroads: Papers Presented to the International Congress of the History of Science and Technology, London: Cass & Co. 1971, with Foreword by Joseph Needham and introduction by Gary Werskey).

Hughes, T. P. 1989. *American Genesis: A Century of Invention and Technological Enthusiams.* New York: Viking.

Holton, G. 1993. *Science and Anti-Science.* Cambridge, MA: Harvard University Press.

Huxley, J. 1930. *A Scientist among the Soviets.* London: Chatto & Windus.

Jacob, M. 1988. *The Cultural Meaning of the Scientific Revolution.* New York: McGraw-Hill.

Johnson, J. 1990. *The Kaiser's Chemists. Science and Modernization in Imperial Germany.* Chapel Hill: The University of North Carolina Press.

Kevles, D. 1987. *The Physicists.* Cambridge, MA: Harvard University Press.

Lederman, Leon. 1991. "The End of the Frontier." *Science* 251 (January Supplement):3–20.

Leslie, S. 1993. *The Cold War and American Science. The Military-Industrial-Academic Complex at MIT and Stanford.* New York: Columbia University Press.

———. 1994. "Science and Politics in Cold War America. Pp. 199–233 in Jacob, M. (ed.), *The Politics of Western Science, 1640–1990.* Atlantic Highlands, N.J.: Humanities Press.

Luttwak, E. 1994. *The Endangered American Dream.* New York: Simon and Schuster.

Mendelsohn, E. 1994. "The Politics of Pessimism: Science and Technology circa 1968." Pp. 151–173 in Ezrahi, Y., Mendelsohn, E., and Segal, H. (eds.), *Technology, Pessimism, and Postmodernism.* Amherst: University of Amherst Press.

Price, D. 1963. *Little Science, Big Science.* New York: Columbia University Press.

———. 1964. "The Science of Science." Pp. 195–208 in Goldsmith, M., and Mackay, A. (eds.), *The Science of Science.* Harmondsworth: Penguin Books.

Rydell, R. 1993. *World of Fairs.* Chigaco: University of Chicago Press.

Schmidt-Pauli, E. von. 1943. *Friedrich Bergius: Ein Deutscher Erfinder kämpft gegen die Englische Blokade.* Berlin.

Shinn, T. 1994. "Science, Toqueville, and the State: The Organization of Knowledge in Modern France." Pp. 47–80 in Margaret Jacob (ed.), *The Politics of Western Science, 1640–1990.* Atlantic Highlands, N.J.: Humanities Press.

Theocharis T., and Psimopoulos, M. 1987. "Where Science Has Gone Wrong." *Nature* 329 (15 October):595–598.

Vucinich, A. 1984. *Empire of Knowledge . The Academy of Sciences of the USSR (1917–1970).* Berkeley, CA: University of California Press.

Werskey, G. 1978. *The Visible College. The Collective Biography of British Scientific Socialists of the 1930s*. New York: Holt, Rinehart and Winston.

Yergin, D. 1977. *Shattered Peace: The Origins of the Cold War and the National Security State*. Boston: Houghton Mifflin Co.

Ziman, J. 1994. *Prometheus Bound. Science in a Dynamic Steady State*. Cambridge: Cambridge University Press.

CHAPTER 7

Postacademic Science

Constructing Knowledge with Networks and Norms

JOHN ZIMAN

EDITOR'S INTRODUCTION

Where is science going after the tapering off of the growth curve of post–World War II Big Science? John Ziman, a theoretical physicist turned theorist of science, has reflected on the fate of science in many books, among others Prometheus Bound: Science in a Steady State. *In his chapter "Postacademic Science: Constructing Science with Networks and Norms" (first given as a lecture to the members of the Royal Society), Ziman acknowledges the inevitable move toward what he calls "postacademic science." He welcomes many of the changes in scientific organization and culture envisioned by the authors of one of the mid-1990s much discussed books:* The Production of Knowledge. *However, he warns that the move toward such things as scientific networks, local scientific cultures, and new criteria for "good science" tied to more immediate applicability of results, may have an unrecognized backside. The disappearance of traditional academic science with its emphasis on public knowledge and an universalist ambition may have consequences that reach beyond science itself. Ziman argues that academic science has come to represent an* important model for society as an example of fact-based dialogue. *A change toward a new criterion for objectivity in postacademic science—however scientifically sound and practically useful—may have unwanted consequences for democracy itself. In other words, Ziman addresses the very same theme as those who worry about antiscience: science as the carrier of crucial* cultural *values. However, he locates the threat not in "antiscience" but in the changing patronage of science itself.*

SCIENCE AS A CULTURAL FORM

A lot has happened to academic science since Peter Medawar taught us to see research as the "Art of the Soluble," nearly thirty years ago (Medawar,

1967). So much has changed that a quite new research culture—"postacademic science"—is emerging. This chapter is concerned with the philosophical impact of these changes.

Let me make it clear that I am not suggesting that present-day scientific results might be less secure, or less in accordance with the true nature of things, than they were thirty years ago. But even the strictest realist would agree that the progressive unveiling of nature is not a very systematic process. How far we have got in that process—that is, what counts as scientific knowledge at any given moment—is obviously influenced by how research is organized, who is involved in it, what they think they are doing, what is regarded as good work, and other similar considerations. In other words, some aspects of the philosophy of science cannot be disentangled from certain features of the current research culture.

ACADEMIC SCIENCE

Academic science emerged in France and Germany in the first half of the nineteenth century. Since then it has evolved into a characteristic social activity, and spread around the world. As the name suggests, it is typically associated with higher education, but is also found in a number of other institutional settings, especially under governmental patronage. It does not have any system of overall control, and although its practices and principles are remarkably uniform, they are not formally codified. For this reason, it is best thought of as a culture, in the anthropological sense, rather than as an organized structure.

One of the questions that can be asked about a culture is whether its practices, rules, traditions, and conventions can be related to a set of more general principles. In 1942, Robert Merton (1942/1973) suggested that academic science was governed by an "ethos" embodying a set of functional "norms," This type of sociological analysis is now considered very questionable, but it does provide useful pegs on which to hang a general account of some familiar social characteristics of academic science, and to relate them to some well-known cognitive features of scientific knowledge.

ELEMENTS OF THE SCIENTIFIC ETHOS

Let us go through the Mertonian norms one by one. The norm of communalism requires that the fruits of academic science should be regarded as "public knowledge." It thus covers the multitude of practices involved in the communication of research results to other scientists, to students, and to society at large. It is no accident, for example, that academic science is closely associated with higher education, or that academic scientists are concerned about the inadequacy of "public understanding of science."

This norm has a deeper significance. In effect, it enjoins the pooling of personal knowledge gained from individual experience. From this shared experience we infer the existence of an external world, on some of whose features we find that we agree. Despite all the arguments of doubting philosophers, scientists are instinctive realists—like most ordinary mortals.

The norm of universality requires that contributions to science should not be excluded because of nationality, religion, social status, or other irrelevant criteria. In practice, this multicultural ideal is achieved very imperfectly. It does imply, however, that scientific propositions should be general enough to apply in any cultural context. This norm thus explains why philosophers of science focus on fundamental theories that claim to reduce and unify a wide variety of phenomena.

The notion that academic scientists have to be disinterested seems to contradict all our experience of the research world. What it means is that in presenting their work publicly they must repress their natural enthusiasm for their own ideas, and adopt a neutral, impersonal stance. Many academic scientists do not have to boost themselves to make a living, because they hold permanent posts as university teachers, and undertake "pure" research without commercial applications. This is a very important norm, since it underpins the philosophical objectivity of academic science.

Originality energizes the scientific enterprise. Academic scientists are not always inspired by curiosity, but they are expected to be "self-winding" in their choice of research problems and techniques. Their most cherished traditions celebrate and sustain this aspect of academic freedom. This is the norm that keeps academic science progressive, and open to novelty. For example, many philosophers of science stress the creative role of conjectures, that is, bold thrusts of intellectual originality, continually attacking the frontiers of ignorance.

Skepticism, finally, is the normative basis for many academic practices, such as carefully controlled critical controversy and peer review. This norm is not a licence for systematic philosophical doubt, nor for total sociological relativism. It merely stresses the constructive role of refutation as the natural partner of conjecture in the production of reliable knowledge. This social mechanism thus tests the claims of academic science in terms of rational qualities such as logical and factual consistency.

The concept of a definite "scientific method" is now considered highly questionable. The most that metascientists will say nowadays is that science is a body of knowledge "regulated" by certain general principles.

These principles are usually considered quite abstract and "philosophical." What I have tried to show, is that they are closely connected with various "sociological" features of academic research.

CUDOS INSTITUTIONALIZED

There is much more to the practice of academic science than individual activity guided by a general ethos. Even the loneliest "seeker after truth" must eventually interact with other people, if only as informed critics or supporters. Academic science could not function without some sort of internal social structure.

This structure is provided by subject specialization. Academic science is divided into disciplines, each of which is a recognized domain of organized teaching and research. It is practically impossible to be an academic scientist without locating oneself initially in an established discipline. The fact that disciplines are usually very loosely organized does not make them ineffective. An academic discipline is much more than a conglomerate of university departments, learned societies, and scientific journals. It is an "invisible college," whose members share a particular research tradition. This is where academic scientists acquire the various theoretical paradigms, codes of practice, and technical methods that are considered "good science" in their particular disciplines.

Specialization does not stop there. The subdivision of disciplines into very narrow research specialties seems to be an unavoidable feature of academic science (Ziman, 1987). But this reveals contradictions in the academic ethos. Most academic scientists can only satisfy the norms of originality and skepticism by concentrating for years on what is known, what is conjectured, and what might be feasible, in a limited "problem area." Unfortunately, this is often inverted into a pedantically sterile or fashionably conformist ideal of "truth." Excessive specialization also encourages sectarianism and cognitive fragmentation, thus offending against the norms of communalism and universalism.

A recognized discipline or subdiscipline provides an academic scientist with a home base, a tribal identity, a social stage on which to perform as a researcher. The academic ethos says nothing about individual motivation. Note, however, that the Mertonian norms combine into the acronym CUDOS—that is, "acclaim," or "prestige." The argument is that academic scientists undertake research, and make public their findings, in exchange for "recognition" by their peers. The citations in the literature, prizes and medals, exalted titles, and other tokens of communal esteem are not just frippery: they are important functional elements of the academic culture.

One question, however, remains unanswered: How do academic scientists make a living? The academic ethos seems to take it for granted that research is a personal vocation, rather than gainful employment. Academic scientists are often deeply committed to their work, but they are seldom "amateurs" in that sense. The peculiar feature of academic science is that it developed as an activity engaged in principally by "academics," whose official employment is to teach, rather than to do research. Everybody knows, of course, that university teachers usually owe their posts to their proven research competence, and earn further promotion by their research achievements. Nevertheless, the convention is that this research is "their own work," which they are free to undertake and benefit from entirely as individuals.

The existence of academic science as a distinctive cultural form thus depends on the willingness of universities and other institutions to provide personal time and other resources for an activity from which they do not directly profit and that they do not directly control. This applies particularly to bodies that support full-time researchers on essentially the same terms, regardless of whether they perform other services, or even whether their contributions to knowledge are of any great significance. This is not the occasion for a discussion of the benefits that academic science brings to society, nor for a demonstration that these benefits far outweigh their out-of-pocket costs. The key point is that academic science relies on public and private *patronage*, in the broadest sense of that old-fashioned word. Its whole ethos is based upon the belief that the pursuit of knowledge is of value in itself, and that only very knowledgeable people can be trusted to pursue it effectively.

NEW MODES OF KNOWLEDGE PRODUCTION

Much more could be said about academic science as an "implementation structure" for the pursuit of a common purpose by a community of individuals activated by a shared culture. This very schematic account does show, however, the linkages between the main features of the culture and some important characteristics of scientific knowledge. Many questions about these linkages remain to be answered. We know for, example, that scientists seldom abide perfectly by their norms: Does that lead to serious imperfections in the output of their research? In any case, should not academic science be considered to be just a subculture of society at large? What are its relationships with other knowledge-based subcultures, such as industrial research or technological practice?

Unfortunately, academic science is changing so rapidly that the real question is What is taking its place? Some of these changes simply reflect scientific and technological progress. As always, the dedication of science to originality is drawing it into quite novel modes of activity. Individual achievement is being merged into the collective action of multidisciplinary teams. Communication is being speeded up electronically, until it becomes instantly global. Instrumental sophistication is making it much easier, but much more expensive, to do good science. Although these may look like natural technical developments, they involve radical changes in many traditional practices and attitudes.

In a recent work (Ziman, 1994), I concentrated on the forces pressing on academic science from society at large. In effect, the whole enterprise, having grown steadily for centuries, has now become too large and expensive to be allowed to go its own way. The governments that mainly fund academic research are putting strict financial ceilings on their patronage, and are trying to get better value for their money. They are insisting that researchers should become much more accountable, more responsive to societal needs, more directly concerned about the quality and impact of their products, and so on. The transition to a "steady state" regime is thus imposing on academic science a number of requirements that are quite foreign to its ethos, and thereby transforming it as a cultural form.

This metamorphosis is still going on. I have been reluctant, personally, to speculate on what will eventually emerge from the present jumble. But now six very distinguished metascientists—let me call them the GLNSST group, for short—have boldly presented a credible scenario for the future of science (Gibbons et al., 1994). In sum, they argue that the academic mode of knowledge production is being replaced by a very different activity, which they call Mode 2.

To some extent the GLNSST argument merely extrapolates current trends within academic science, such as the growth of multidisciplinary teamwork and interinstitutional networking. But it also suggests a decisive break with the academic tradition in relation to conditions of employment, problem choice, criteria of success and other important features. In other words, Mode 2 is not just a "new mode of knowledge production": it is a formula for a possible new research culture.

The GLNSST group note that Mode 2 has evolved outside academia, and will not necessarily supersede Mode 1 in its traditional setting. But this is a very real possibility, which ought to be taken seriously. At least it provides a coherent model for "postacademic science." Following a methodological principle that I learned as a theoretical physicist, I shall explore this model as a cultural form, and ask what sort of knowledge it might be expected to produce.

NETWORKING INTELLECTUAL PROPERTY

Although GLNSST present Mode 2 as a coherent activity in its own right, they describe it mostly in terms of its differences from Mode 1. For this reason, Merton's list of norms again provides a convenient analytical frame. We start, then, with the norm of communalism. This looks easy. Just as "community" was the keynote of academic science, so "network" has become the token of postacademic science. In both cases, science is visualized as a communication system, where information obtained at certain nodes is transmitted to other nodes, whether these be individual researchers, research groups, specialist communities, corporate bodies, or the general public.

Nevertheless, certain technical and procedural developments have produced significant cultural and cognitive effects. The increasing density, multiple-connectivity and immediacy of electronic communication draws individual researchers together into collective action. This is not just a matter of facilitating active teamwork by geographically dispersed researchers. It makes it feasible for novel observations and theories to be discussed in detail with distant colleagues—or even skeptical rivals—as they emerge. Databases and archives can be searched thoroughly for relevant ideas and information. An electronic text can be amended so easily that it need not come to a firm conclusion until this has become acceptable beyond refutation. The various phases of the research cycle—discovery, justification, criticism, and revision—merge together in an off-the-record process involving a whole cluster of informal contributions. The material that does eventually get into the official scientific literature may thus already represent a wider consensus than the group of authors to whom it is officially attributed, and should therefore be less tentative, more convincing, sounder in fact and logic, than is normal in academic science.

On the other hand, Mode 2 networks are typically very heterogeneous. Academic scientists are regularly teamed up with researchers who are not bound by the norm of communalism, and are not professionally dependent on their contributions to "public knowledge." Mode 2 knowledge is proprietary. Research results that an academic scientist would have published as soon as possible are now defined as "intellectual property," and may be kept secret for longer or shorter times for commercial reasons. This has the effect of weakening the traditional mechanisms motivating prompt publication. It also means that the knowledge appearing in public out of postacademic science may lack significant items that are only known to a privileged group, such as the employees of a particular industrial firm.

PROBLEM SOLVING IN LOCAL CONTEXTS

From a sociological point of view, Mode 2 fully satisfies the norm of universalism. The networks of communication and collaboration are global. It is not necessary for a researcher to move bodily to an established research center to do good science. Researchers located in industrial firms, government laboratories, charitable foundations, and universities may work together in the same team. Even the tribal boundaries between disciplines are disregarded. As GLNSST point out, this universalism is not a remedy for gross inequalities of resources, facilities or competence between countries or institutions: but that is a much larger issue than can be considered here.

On the other hand, postacademic science may not favor metaphysical universalism. Mode 2 evolved outside academia, as a technique for applying science to practical matters. It is organized intellectually around the solution of problems, rather than directed toward the production of knowledge as such. It follows that the knowledge that is actually produced is intrinsically local, rather than universal. Even though it may have wide theoretical implications, it is not shaped by a preference for unification and generality. What counts as "good science" in Mode 2 may be technical skill at problem solving, rather than advancement of our understanding of the natural world.

This does not mean that postacademic science must necessarily be "useful." It should not be thought of as inverting the academic tradition of "purity" and "uselessness." Utility will simply cease to be a demarcation criterion. One of the main features of Mode 2 is that it draws on, and generates, problems, techniques, and research results from all parts of the conventional "R&D spectrum." Basic research and technological development already interpenetrate one another: in the long run, they will become inseparable.

Instead of unification postacademic science favors finalization (Böhme et al., 1983). Like most general terms used to describe the research process, this is an ill-defined concept, with ambivalent policy resonances. But it indicates the way that research in a particular area may become more "mature," until it is guided by an unchallenged theoretical paradigm. Research programmes are then be formulated within a stable framework of concepts and techniques, and directed toward specific ends. "The art of the soluble," as practised intuitively by individual researchers, gives way to an explicit rationale for the way in which problems are posed and attacked.

Finalized research is not free of uncertainty, and its objectives are not necessarily utilitarian. The orientation of Mode 2 toward specific problems actually permits a looser, more experimental approach, where the problem itself provides the focus of continued effort. This approach works well in the study of natural and artificial phenomena in problem areas whose contexts are partly universal and partly local. But it is also perfectly capable of taking on the well-posed basic problems that arise naturally in the regions where traditional disciplines interact or overlap.

Finalization favors interdisciplinary research, since it obviously benefits from the reduction of observed phenomena to more fundamental principles. But postacademic science will probably not be driven by reductionism as a metaphysical ideal. A striving for "local" understanding, without preconceived notions of what might require explaining or be acceptable as an explanation, may even be more effective in closing the gaps in the knowledge map than a single-minded pursuit of general intellectual unity.

INCORPORATING INTERESTS INTO KNOWLEDGE

The norm of "disinterestedness" was always difficult to sustain. Even university teachers engaged in "pure" research have strong professional interests, and are not completely shielded from economic and political pressures. In Mode 2, knowledge is produced by teams of researchers networked across a wide range of organizations. The diverse working conditions of these researchers cannot be disregarded. Those who are on short term contracts of employment are not in a strong position to show independence of mind. Those who are employees of industrial firms must always be aware of the potential commercial value of what they produce. In the latter case, the suspicion of bias cannot be entirely discounted just because the research seems disconnected from any possible application. An industrial firm is not a charity. If it does not think it will benefit in some way from the work, then how can it justify paying its employees to do it? Postacademic science will surely be too deeply entangled in networks of practice to be considered free from such influences. For most sociologists and philosophers of science, this is not a new thought. The notion of a truly objective, disinterested "seeker after truth" is not consistent with the realities of social existence. We all have interests and values that we are bound to promote in our scientific work, however hard we try to suppress them.

But the actual effect of these factors can be exaggerated. The essence of the academic ethos is that it defines a culture designed to keep them as far as possible under control. Academic science does often manage to live almost up to its ideals. Mode 2, by contrast, does not just "produce" knowledge: it is a culture where knowledge is constructed in accord with the commercial, political, or other social interests of the bodies that underwrite its production. Although these interests may also incorporate scientific values, this is a culture where socioeconomic power is the final authority. If that is indeed the way that postacademic science is going, then the sociological relativists who insist that scientific knowledge always serves such power will be proven right after all.

WHO SETS THE PROBLEM?

Mode 2 research is performed in a "problem context." This is nothing new in principle. Philosophers and scientists agree that the identification of a significant but soluble problem is a vital stage in the research process. The question is Who sets the problems? Academic science works on the assumption that researchers are free—within reasonable limits—to set their own problems. In Mode 2, by contrast, researchers work together on problems which they have not posed personally, and which they may not even have chosen collectively as a team.

According to GLNSST, Mode 2 problems are set in "hybrid fora" where the researchers involved may have more or less weight, depending principally on how close the problems are to practical application. But even basic research does not take place in a power vacuum. It has to be supported financially and administratively by bodies whose interests go beyond the mere production of knowledge. They naturally exercise these interests at the point of maximum leverage—that is when research problems are being set. All policy talk about foresight, priorities, accountability, and so on is really focused on "problem choice."

Postacademic science will surely not be given over completely to "commissioned" research. It will be on the lookout for technical virtuosity by individuals and research teams. But Mode 2 tends to define the highest form of scientific "creativity"—the construction of soluble research problems—as a group phenomenon. This is why academic scientists set such great store by "responsive mode" funding, which provides them with almost their only opportunities to demonstrate this attribute as individuals. Even then, success in formulating fundable research proposals may simply reflect a shrewd eye for what is currently regarded as "good science."

Academic science has always worked on "Darwinian" principles. Scientists undertake research and offer results on innumerable different problems. Knowledge advances in unforeseen directions by the retention of the small proportion of these results that survive rigorous testing. It is a very wasteful process in detail, and seldom produces outcomes that perfectly fit our current needs. It just happens to be extraordinarily effective at generating remarkably reliable knowledge.

Postacademic science will continually strive to improve on this process. It will try to push the process in desired directions by strongly favoring research on particular problems. It may thus be very effective in overrunning disciplinary frontiers, in order to construct detailed local maps of potentially useful areas.

Postacademic science will also try to eliminate waste by ensuring that all research projects are well-designed, and directed toward well-posed problems. But the efficiency of a Darwinian process depends on having a highly diversified stock of variants, as well as a highly selective environment. The effect of "collectivizing" problem choice must be to limit the range of variation of research projects. Mode 2 problems are socially preselected, on hypothetical grounds, before they are actually tackled. This may mean that the science that gets done is "better" and more "relevant" than if it were left entirely to the idiosyncratic judgments of individual scientists. But it also means that a few wild conjectures never get a chance to show their hidden capabilities, which are just occasionally revolutionary.

WHAT COUNTS AS EXCELLENCE?

Mode 2 replaces peer review of research outcomes by quality control of people, projects, and performance. But this usually embodies a much broader notion of "excellence" than the traditional academic criteria for "good science." Proven competence as a researcher may count for less than a good record as an expert at solving societal, environmental, or commercial problems. This expertise need not be very specialized. It may be just an ability to enter a temporary research team smoothly, and make a useful contribution. The research quality of a team may be confused with its success in getting funding. More and more importance may be attached to entrepreneurial and managerial skills as the research process becomes part of a larger cycle of action—for example, in the successive stages of bringing a technoscientific innovation to market.

Paradoxically, postacademic science could become so obsessed with accountability, performance monitoring, contractual scrutiny, and other forms of "quality control" that it sacrifices the quality of the procedures themselves to their sheer quantity. Mode 2 research does not promote the establishment of groups of practitioners in stable positions of intellectual authority. In the absence of human reference groups, assessment procedures may be automated. Quality control is then made to rely on surrogate indicators of performance, whose legitimacy may well be questioned on scientific grounds.

In other words, mode 2 downplays the role of systematic intellectual criticism, which is the key to the validity of academic science. In contexts of application, practical utility must eventually be effective as a selection mechanism, even if only in pragmatic terms. But in fundamental research, where this mechanism cannot operate, what Merton called organized skepticism is the only real protection against the embodiment of serious errors in the knowledge that is produced. Perhaps a higher level of cognitive insecurity is a price that will have to be paid as postacademic science becomes more entangled with "trans-epistemic" issues, involving societal, environmental, and humanistic values.

FROM SPECIALIZED KNOWLEDGE TO EXPERTISE

The world of practice does not carve itself up neatly along the joints between the academic disciplines. In the context of application, all problems require a multidisciplinary approach. This is equally true for research into fundamentals, where the most interesting areas for the exercise of intellectual curiosity are interdisciplinary. The most radical feature of Mode 2 is that it strives to take a broader view than can be achieved from within any one discipline. In Mode 2, specialists from different disciplines work together as a team. Ideally, their different approaches to the problem fuse together into a coherent attack with a comprehensive solution. Between them, they construct a transdisciplinary schema of paradigms, techniques, and expertise that provides a framework for further advances. In traditional academia, one would say that a new specialty was emerging, and expect to see it institutionalized as a regular discipline. But Mode 2 is not geared for such a development. There is no organizational or intellectual structure on to which a research team can crystallize and exploit its transdisciplinary capabilities. After a certain time, the participants must break away and form new configurations, around new problems, requiring a different mix of skills.

The GLNSST vision of free-floating, mutable, transdisciplinary schemas lacks a social context. In reality, practical problems seldom appear out of nowhere, without antecedents. The world where research is to be applied is already highly structured. That is, the problems to be tackled will normally be set and funded by their organizational "owners," such as industrial firms, government departments, health services, and so forth. In the effort to overcome the academic vice of narrow specialization, postacademic science may find that it has put itself into the hands of bodies that are even more parochial, fragmented, and restrictive than the disciplines from which it has escaped.

WHERE DO THE PIPERS COLLECT THEIR PAYCHECKS?

Mode 2 researchers work in shifting teams, like small firms producing goods for a competitive market. This system attaches the individualism of academic science to small groups, and motivates them with entrepreneurial insecurity. This insecurity is endemic. Even if such a group is not an independent entity, always risking its future in the competition for funding, it cannot provide stable employment for most of its members. As teams reconfigure to tackle new problems, some researchers will have to move elsewhere to make room for new people with new skills. Academic scientists are often demoralized by undue persistence in tenured, specialized research. Mobile researchers hired as professional problem-solvers may be equally demoralized by lack of stable opportunities to establish or exercise their specialized expertise.

The Mode 2 model of an open system of economically independent entrepreneurial groups also assumes the existence of a market for research services and results. Markets of this kind do exist in certain high technology industries, although close study shows that small research enterprises survive there mainly through their connections with very large commercial firms. Government funding of more basic research is often organized around competitive project grants, customer-contractor arrangements and other market concepts. But the entities that actually compete in the provision of these services are seldom freestanding. The researchers who network across the world are mostly full-time employees of universities, government laboratories, charitable foundations, or industrial firms. They may have to fight to keep their jobs, but they do not have to take financial responsibility for the elaborate facilities that they actually use in their research.

Mode 2 research looks attractively unbureaucratic, but it is really heavily capital-intensive. It is funded by a complex of governmental bodies, large public institutions, and private corporations. It could not exist without this economic base. Postacademic science will surely not be able to duck the central questions of science policy: Who will pay the pipers, and what tunes should they be called on to play?

POSTINDUSTRIAL SCIENCE

The new mode of knowledge production described by GLNSST is a very different culture from Mode 1. In fact, it evolved in a different social niche. The systematic use of scientific research to solve practical problems is at least as old as academic science. Medicine, engineering, agriculture, mining, and particularly manufacturing industry, have all nurtured applied science, and benefited immensely from it. This is where Mode 2 came from, and where it still largely fits.

Applied science has expanded so rapidly and diffused so widely that it now greatly exceeds academic science in scope and scale. But it never developed a homogeneous culture. It is distributed in pockets throughout society, and in each pocket it is shaped to fit local practices. Clinical medicine, for example, is a very different profession from engineering, and organizes its research activities very differently.

Nevertheless, the industrial applications of science have been so dramatic that applied science is often identified with the research and development activities of large commercial firms. By the 1960s, industrial science had emerged as a characteristic way of life.

Technically and cognitively, industrial science was the twin of academic science, and yet it was organized on quite different principles. Indeed, these principles are almost the inverse of the Mertonian norms (Ziman, 1995). The standard form of industrial science was Proprietary, Local, Authoritarian, Commissioned, and Expert. I have not time here to go through these characteristics one by one, but it is no accident that they spell out PLACE. The stereotypical "industrial R&D laboratory" offered a lifelong organizational career in a hierarchy of technical specialties. In effect, it was a managerial microcosm of the industrial firm in which it was embedded.

Since that time, however, industry itself has changed. We are said to be entering a "postindustrial" era, characterized by multinational firms, decentralized managerially into small, specialized service units, devolving much work to subcontractors, coordinated globally by information technology, and so on. And as industrial firms changed their working methods, they restructured their research activities along similar lines. Their R&D laboratories were devolved into multidisciplinary matrices and global networks of temporary project teams, buying in specialist functions from independent contractors, and so forth.

In a word, Mode 2 is essentially the "postindustrial" version of applied science. Postindustrial science differs from earlier forms by substituting "market" competition for "command" management, but is actually based on the same principles. As we have seen, it is "proprietary," "local," "authoritarian," "commissioned," and "expert," even if it does not offer such a safe "PLACE" as it used to.

The differences between Mode 1 and Mode 2 are not just signs of a recent change from an old to a new method of knowledge production. They have their roots in the historical distinction between pure and applied research—a distinction that was embodied institutionally in the gap between academic science carried out in universities and industrial science carried out in industrial laboratories. This gap corresponded to a real cultural difference between two social systems. But these systems were always closely connected and dependent on one another. They could not live easily together under the same roof, but the existence of each was always understood to be essential to the continued vitality of the other.

The evolution of industrial science into postindustrial science is altering this relationship. We have already identified a number of factors working toward a single "postacademic" culture. Cognitive developments, such as the "finalization" of many subdisciplines, are blurring the distinctions between basic and applied research. Technological developments, such as real-time electronic networking, are generating heterogeneous, hybrid teams that override institutional loyalties. Economic conditions, such as the transition to "steady state" funding, are forcing the two cultures into the same organizational mould.

Indeed, a self-conscious effort at a high level of political and managerial authority would now be required to keep the two systems from coalescing in style and function. But such a merger not only raises many practical issues of funding, disciplinary identity, criteria of excellence, career aspirations, intellectual property rights, institutional management, and so on. It also brings face to face two very different sets of structural principles.

We are thus faced by the classic sociological situation discussed nearly a century ago by Max Weber. In any such confrontation, the organizational prescripts of PLACE will almost certainly prevail over the communal norms of CUDOS. In other words, as GLNSST surmise, Mode 2 will largely, if not entirely, supersede Mode 1 throughout the world of science. The culture of postacademic science will be predominantly postindustrial.

POSTMODERN KNOWLEDGE

The transition from academic to postacademic science will surely leave the operating philosophy of research unchanged. Scientists will still construct knowledge on the basis of a firm belief in the existence of an external world whose behavior is intelligibly regular and not disjoint. They will go on theorizing, and testing their theories by observation and experiment. They will continue—quite rightly—to resist firmly the philosophical skepticism, sociological relativism, political cynicism, ethical nihilism, and historical incommensurabilism projected on to science by some of its wilder critics. It is not academic science, but academic metascience that is in a state of intellectual anarchy, where "anything goes." Nevertheless, in its postacademic form, science will have shed some of the doctrines of "modernism." In particular, it will not claim to be able to produce a universally applicable answer to every problem. In wider cultural and humanistic circles, "modernism" has now given way to "postmodernism" (Toulmin, 1992). Postacademic science will not only be postindustrial in its social role, and "hypermodern" in its conspicuous engagement with information technology: it will also be postmodern in its philosophy.

Of course, terms such as "modernism" and "postmodernism" are very ill-defined. To most scientists they sound like cult slogans, deployed for, or against, the most diverse fashions and fads. I believe they have a serious core of meaning, but would not pretend to be able to define it. I am using them simply to indicate that there have been changes in the philosophy of science, and that these are not unrelated to changes in our general philosophical outlook.

The new wave that has swept through the nonscientific intelligentsia in recent years does not seem to be a primary cause, or a primary effect, of what has been happening in science. But there are undoubtedly numerous secondary interactions that may not be insignificant. Some of the likely features of postacademic science do resonate with certain elements of the postmodern critique. In pointing out these resonances, I am simply following a fruitful precept of theoretical physics. As the novelist E. M. Forster put it: "Only connect!"—or was it "Where angels fear to tread . . ."?

EPISTEMIC FEATURES OF MODE 2

Let us look again at some of the typical features of Mode 2. In the first place, Mode 2 is not activated by the vision of a unified, universal scientific world picture and does not try to "reduce" every body of knowledge to one that is more "fundamental." This is in line with the postmodern critique of grand theory. Postmodern philosophers renounce the age-old attempt to put human understanding on absolutely firm "foundations." They argue cogently that reductionist explanations of natural phenomena can never be complete or perfect, and that "metanarratives" can never be grounded in absolute, accurate, ultimate truths. Postacademic science will no longer promote the intellectual imperialism of scientific monism.

Mode 2 focuses, rather, on the regions around specific problems. The dense communication networks and transdisciplinary openness of postacademic science will encourage thorough exploration of limited areas, and of the pathways between them. This does not preclude the study of fundamental questions. But instead of advancing head on against conceptual enigmas, research will start with concrete problems and open out in the direction of greater generality.

But as postmodern philosophers point out, the construction of a reliable representation of a local reality usually involves the development of a correspondingly local language. The more elaborate this representation, the more difficult it is to uncouple such a language from its context, and to use it in representing the realities of other problem areas.

As a consequence, postacademic maps of knowledge may well be meticulously detailed and systematic in their coverage, but they will still be divided up into specialized domains, characterized by mutually incomprehensible technical languages. Indeed, these domains will always be in a state of flux, and often overlap one another. Novel solutions will be found for long-standing problems. Novel problems will arise in new contexts of application. Paradigms, techniques, and specialized skills will be continually segmented and recombined into new configurations.

This does not mean that postacademic science will reject operational realism. On the contrary, research will be rooted in life-world problems. The postmodern critique applies to attempts to represent objects existing outside the mind. Such a representation can be perfectly real as a guide to thoughts and actions, but that does not require it to be complete, timeless or unique. It can always be analyzed more and more precisely, and traced further and further back into wider networks of representation. Postacademic science will be enlarged and enriched by this process of deconstruction—typically toward greater generality and abstraction—without necessarily losing contact with the reality.

Postmodernism thus denies that the representation of any aspect of nature must converge toward a unique map. It repudiates the metaphor of the university as a notional "brain," whose permanent modules are academic faculties and departments, each dealing tidily with its allotted discipline. Highly formalized schemes of thought are to be regarded less as strengthening skeletons than as potential barriers to understanding. Here again, postacademic science will not reject such schemes out of hand, but will take a skeptical attitude toward their claims.

Mode 2 is not merely multidisciplinary: it is almost dogmatically pluralistic. It welcomes diversity, and is not fearful of possible inconsistencies. The knowledge that it produces is not organized around theoretical issues, and is not automatically subject to clear rules of coherence and credibility. It may combine cognitive and noncognitive elements in novel and creative ways—witness cognitive science itself—but it can also be a diffuse—even opaque—mixture of theory and practice, ideas and data, designed to meet the needs of a specific application. In other words, in Mode 2 pragmatism rules.

Postmodernism similarly celebrates extreme pluralism. It favors wide definitions of knowledge and decentred diversity. It denies the possibility of formulating general rules by which to judge the validity of new ideas, or stable categories into which to place new data. This philosophical doubt is, of course, mostly—well, "philosophical." Postacademic science need not cast itself adrift from its twin anchors of rationality and empiricism. In its traditional sphere of basic research it will surely maintain the critical apparatus of academic science. But it may become incoherently (and unreliably) postmodern in other spheres, where it forms hybrids with research cultures that do not share the same intellectual values, or the same standards of "good science."

Finally—but perhaps most important—Mode 2 is permeated with social interests. In postmodern terms, it is consciously reflexive. The interaction between the knower and what is to be known is seen to be an essential element of the knowledge. Some allowance can be made for this effect, but it cannot be eliminated.

The parallel with some accounts of quantum theory is obvious, but not relevant. The real point is that postacademic science will always have at least half an eye on the contexts of application from which it gets its problems. It will therefore be dealing with matters where societal values—safety, profitability, efficacy, and so on—cannot be ignored. These values conflict with one another. Individual citizens and independent social groups rightly weight them differently. However hard people try to rise above their personal attitudes, these become significant factors in the cycle of research and action.

Doctrinaire postmodernism deconstructs scientific objectivity out of existence. Postacademic science will surely defend objectivity as an ideal, impossible to realize completely in practice but always to be respected and desired. But if all research arises in contexts of application, there may never be any occasions where this ideal is paramount. Scientific objectivity is not an abstract philosophical virtue. It is a cultural norm embodied in a web of social practices. Academic scientists internalize the norm of "disinterestedness" through experience in research situations where these practices are systematically observed. It is hard to see how this norm will be sustained when there are few situations yielding the relevant experience.

WHAT PRICE OBJECTIVITY?

This analysis does not suggest that science is "going postmodern" in the fullest sense. Most of the postmodern features of postacademic science are quite mild, and even benign. Some are much-needed corrections to the excesses of "scientism." Others are welcome antidotes to the decontextualized rationalism that has long plagued the philosophy of science. Other features, again, help rescue the scientific imagination from entrenched specialization. And localized pragmatism will largely compensate for the fragmentation of theoretical standards of scientific validity.

The decline of objectivity, however, is a much more serious matter. It illustrates perfectly the connection between a cultural norm and a philosophical concept. In the ethos of academic science, the sociological norm of disinterestedness is linked to the regulative principle of objectivity, each reinforcing the other. But postindustrial research has no place for disinterested practices, and postmodern thought has no place for objective ideals. Postacademic scientists will have neither examples of disinterested behavior to emulate, nor formal standards of objectivity to live up to. Constructive reinforcement will give way to deconstructive decay.

Objectivity is one of the features that make science so valuable in society. It is the public guarantee of reliable, disinterested knowledge. Science plays a unique role in settling factual disputes. This is not because it is particularly rational or because it necessarily embodies the truth: it is because it has a well-deserved reputation for impartiality on material issues. The complex fabric of democratic society is held together by trust in this objectivity, exercised openly by scientific experts. Without science as an independent arbiter, many social conflicts could only be resolved by reference to political authority or by a direct appeal to force. That was the historical experience out of which scientific institutions such as the Royal Society emerged, and its wisdom remains our cornerstone.

The postmodern critics of science insist that its claims to objectivity are false, and actually conceal powerful established interests. It is noteworthy, however, that anti-establishment groups also appeal to objective scientific knowledge to advance their causes—for example in environmental disputes. Even metascientific anarchists should realize that by "unmasking" the "ideology" of objectivity they are breaking their own swords in the struggle against their most feared opponents—the corporate and governmental enterprises that drive postindustrial science.

Is there any way of avoiding this loss? The trouble is that scientific objectivity is an emergent cultural property of academic science. It was not an attribute of any previous knowledge-producing culture, and we have no theoretical models for other cultures with similar attributes. The best that we can do is to determine the functions that it serves in academic science, and the features that sustain it there. We can then ask whether these features could be maintained in postacademic science, or at least in some segment of it.

We thus return to the question whether postacademic science is bound to be completely "postindustrial." The GLNSST group define Mode 2 in that way, and surmise that it will supersede Mode 1, even within the university. This surmise is much more significant than it appears at first sight. In particular, the suggestion that all research will eventually relate to problems arising in the context of application has profound consequences. Combined with the demand for accountability in the formulation of problems, it opens up the whole of academic research to the influence of external interests.

We only have experience of one way of countering this trend. But that would take us back to the central tradition of academic science. This was not an ethos or a bundle of communal practices. It was, quite simply, patronage. It was the convention by which society provided resources for the production of knowledge without insisting that they should be accounted for, in prospect or retrospect, in utilitarian terms. This convention may now seem elitist, irresponsible, and inefficient, but it worked remarkably well in its time. We abandon it at our peril.

ACKNOWLEDGMENTS

This chapter is the substance of the 1995 Medawar Lecture of the Royal Society, delivered in London on 29 June, and in Edinburgh on 27 September 1995. I am grateful for helpful comments from Phoebe Isard and Helga Nowotny.

REFERENCES

Böhme, G., van den Daele, W., Hohlfeld, R., Krohn, W., and W. Schäfer. 1983. *Finalization in Science: The Social Orientation of Scientific Progress*. Dordrecht: Reidel.

Gibbons, M., Limoges, C., Nowotny, H., Schwartzman, S., Scott, P., and M. Trow. 1994. *The New Production of Knowledge: The Dynamics of Science and Research in Contemporary Societies*. London: Sage.

Medawar, P. B. 1967. *The Art of the Soluble*. London: Methuen.

Merton, R. K. 1942/1973. "The Normative Structure of Science", Pp. 267–278 in R. K. Merton, *The Sociology of Science*. Chicago: Chicago University Press.

Toulmin, S. 1992. *Cosmopolis: The Hidden Agenda of Modernity*. Chicago: University of Chicago Press.

Ziman, J. M. 1987. *Knowing Everything about Nothing: Specialization and Change in Scientific Careers*. Cambridge: Cambridge University Press.

Ziman, J. M. 1994. *Prometheus Bound: Science in a Dynamic Steady State*. Cambridge: Cambridge University Press.

Ziman, J. M. 1995. *Of One Mind: The Collectivization of Science*. Woodbury NY: American Institute of Physics.

CHAPTER 8

A Social Theory of Objectivity

STEPHAN FUCHS

EDITOR'S INTRODUCTION

In this chapter Stephan Fuchs, a sociologist and the author of The Professional Quest for Truth, *goes to the root of the opposition between the proscience warriors and "postmodernist" academics of various kinds. What has particularly divided the camps in the Science Wars is the question about* the possibility and desirability of an objectivist epistemology. *Recently many feminists and others have been eagerly arguing for the necessity of "standpoint epistemologies" and the abandonment of the idea of objectivity. The idea of abandoning objectivity has deeply upset the proscience warriors—and most practicing scientists as well. (Indeed, some of Gross and Levitt's fiercest attacks have been just on standpoint feminists.)*

But "objectivity" can mean many things. Fuchs traces the many recent attempts by historians, philosophers, and sociologists of science to "locate" objectivity in various ways. Many new studies typically emphasize the thoroughly social *foundation of objectivity; its dependence on such things as particular communication practices, trust, and the like. Still, unlike postmodernist standpoint arguments, many of these studies do* not *typically dispute the existence of objective practices in science, or science's objectivist intent.*

After demonstrating the untenability of standpoint epistemologies, Fuchs suggests that we abandon epistemological musings and get back to a sociological *analysis. He invokes the sociologist Niklas Luhmann's idea of "science" as representing a particular abstract type of social communication. This type of communication is* exclusively interested in the truth versus falsity of claims, *in eliminating error, and in learning from mistakes. This is what "science" is all about, if we analyze it strictly as a form of communication; anything else is secondary. Thus, like Ziman, also Fuchs emphasizes the social importance of the objectivist "attitude" or intent of science.*

THE MANY MEANINGS OF OBJECTIVITY

"Objectivity" has a variety of contested meanings, most of which have been reassessed in the past ten years or so. Objectivity can, for example, be ascribed to a person's capacity for impartial and disinterested judgment. Sometimes, objectivity is seen as a quality of methods and rules of inquiry that discipline arbitrary and accidental forces impacting on knowledge. In a more specific and technical sense, measures are objective when they covary strongly with each other and across repeated measurements taken independently by several investigators; such measures reliably indicate a theoretical entity. As a property of knowledge, objectivity refers to propositions that capture some independent and external reality. Finally, objectivity can be attributed to social and cultural institutions that are somehow more solid and enduring than individual beliefs.

What these different meanings of objectivity have in common is their "negativity." Originally referring to a positive quality of states in the world, objectivity has, since the Scientific Revolution, become the *absence* of individual, idiosyncratic, accidental, and contingent forces and circumstances. For a long time, the care for this absence has been the exclusive responsibility of philosophers who worship objectivity, together with truth, as the most precious epistemic value of modern science. Philosophers generally agree that objectivity is the distinctive mark of scientific knowledge; disagreements exist over how it is possible, how it can be secured, and for exactly what sorts of propositions objectivity can be achieved. In what follows, I want to *sociologize* objectivity. That is, I deal with objectivity as a special kind of social and communicative "medium" that separates science from other modes of communicating, such as love proposals, legal threats, ideological warfare, and transactions in markets and hierarchies.

Think of scientific knowledge as emerging from the interaction between two forces: Knowers and the known world. Objectivity results when the contributions to knowledge of knowers—be they conceived as transcendental Subjects, empirical minds, or social groups—are reduced or, better still, eliminated in favor of the contributions of the world to that knowledge. In this sense, objectivity is a matter of conduct, of controlling emotions, biases, and interests. Once subjectivity is disciplined, the world can impress itself on knowledge like a seal on wax, without interference from culture or society. Then, there is a good chance that objectivity will result as accurate knowledge of external reality.

As both conduct and knowledge, objectivity rises above partial perspectives and idiosyncratic standpoints, and from there allows for intersubjective consensus as well. The historical semantics of objectivity gradually shifts from "objectivity" as a quality of the world itself, to a property of knowledge about it (Dear, 1995, chapter 1). The knower or observer gradually disappears, is rendered invisible and anonymous, and finally turns into a mechanical and replaceable appendix of the instruments. The climax of this development are self-recording devices, where observers enter only as sources of contingent psychological irregularities, which then can be canceled out by statistical averaging procedures, such as least squares or rules for taming outliers (Swijtink, 1987).

This "aperspectival" (Daston, 1992, p. 599) objectivity can be achieved, it is said, by letting the world select its own representations, with minimal interference and distortion by the interest- and standpoint-driven forces of society, culture, money, and power. In Popperian language, objective knowledge is knowledge without knowing subjects. In Nagel's (1986) famous phrase, it is the "view from nowhere." In Rorty's (1979), no less famous, imagery, objectivity is what the Mirror reflects when all contingencies have been removed.

Taking this view requires special efforts. Objectivity does not come easy or naturally. Rather, it requires the painstaking and patient *work* of removing the biases and prejudices that are part and parcel of the profane world (Dear, 1992, p. 627; Daston, 1991a, p. 353). Left to themselves, without systematic and prolonged disciplining, training, and control by their scientist-peers, knowers cannot be expected to arrive at objectivity. By themselves, they are victims of the "idols." Avoiding bias and distortion requires a special Ethos and virtue; a heroic and ascetic Weberian denial of idiosyncratic selves and their social memberships (Merton, 1973; Daston, 1991b, p. 380; Daston & Galison, 1992, p. 83).

Epistemology has suggested various ways of doing this work, all of which seem to have failed, at least in their strongest foundationalist and absolutist aspirations (cf. Laudan, 1996, pp. 3–25). Some of the most prominent attempts include the Kantian transcendental Subject, the Cartesian Cogito, the neutral and invariant protocol sentences of Logical Positivism, the elusive demarcation criterion separating science as a natural kind from other knowledge, the internal/external distinction between the context of justification and the context of discovery, the ideal speech situation of formal pragmatics, and the rules of scientific method. These devices were all meant to banish or, at least, hold at arm's length all those profane economic or social forces that impacted on knowers to obscure and distort or bias their view. The priestly philosophers erected, and anxiously guarded, taboos protecting the sacred institution of objectivity, where science must only be driven by the internal and pure forces of Reason (logic) and Reality (sensory evidence).

Against this philosophical orthodoxy, the antifoundationalist and antirepresentationalist consensus in science studies points out that science is "just" a contingent and historical culture or form of life—a communal and local language game without any privileged access to objectivity and truth. Scientific knowledge is "constructed" much like other knowledge;

it is local and contextual, maybe "gendered" and politically suspect. The bottom line in current science studies is that nothing special is happening in science. It is an ordinary setting where ordinary people negotiate definitions of situations, and tinker with tools and instruments to achieve practical results. Scientists do not follow some algorithmic and universal "logic of science." As Shapin (1995, p. 305) says in his recent review of the field, science studies display the "contingency, informality, and situatedness of scientific knowledge-making." This is consistent with postempiricist, pragmatist, and constructivist philosophies of science that bid farewell to the old promises of absolute objectivity and foundational certainty (Rorty, 1991).

Two conclusions are frequently drawn from the constructivist critique of objectivity. The first conclusion is that objectivity is rhetoric; the second one holds objectivity to be power.

OBJECTIVITY AS RHETORIC

To view objectivity as rhetoric has become very popular indeed, due to the expansion of semiotics, literary criticism, and discourse analysis into the humanities and social sciences (Fuchs & Ward, 1994). According to this textualist position, objectivity is mainly a set of textual and literary devices to persuade audiences that some statement or proposition accurately represents an independent reality "out there" (see Bazerman, 1988; Woolgar, 1988; Myers, 1993). The rhetoric of objectivity produces an "ideology of representation," where accurate statements, or facts, seem to correspond to external reality (Mulkay, 1985; Hunter, 1990, pp. 5–6).

The appearance of correspondence is accomplished by various rhetorical and "inscription devices"—such as externalization, distention, or the passive voice—that eliminate all references to agency, locale, subjectivity, and construction (Latour & Woolgar, 1979/86, pp. 81–88). Gergen (1992, p. 104) defends this position:

> Linguistic means of separating subject and object, characterizing the objective world, establishing authorial presence (and absence), and cleansing the lens of perception are among the most prominent means of generating the sense of objectivity

Deleting all modalities and qualifiers that would restrict the scope of statements by connecting them to local and accidental conditions, the "empiricist repertoire" persuades audiences and readers that a text indeed mirrors reality and contains objective knowledge (Gilbert & Mulkay, 1984). Gusfield (1976) characterizes scientific prose as the "style of non-style," which encourages readers to believe in the objectivity of what is being communicated. Consequently, "when one either fails to employ distention devices or resorts to personalizing descriptors, objectivity is threatened" (Gergen, 1992, p. 97).

This argument for objectivity as rhetoric has some merits, but overstates its case. Most important, the rhetoricist argument gives almost magical powers to the Word, the Text, and Discourse, while ignoring the organizational and material contexts of scientific work. Rhetoric is but a small part of this work, surfacing, as it does, primarily on the public front stages of scientific presentations. That scientists use rhetoric to persuade others is true but trivial—all organizations, and especially their professional segments, do this extensively (Meyer & Rowan, 1977). The more interesting and less obvious question, though, is when such rhetorically couched claims to objectivity and disinterestedness are likely believed, and when they are met with the suspicious skepticism of *Ideologiekritik—no matter* what rhetorical tropes are being employed by a given text.

Credible claims to objectivity are supported by much more than just rhetoric; they rely on machines, statistics, powerful organizations, professional monopolies, strong networks, and vast amounts of reputational capital (Whitley, 1984). Galison (1985, pp. 356–359) shows how in today's Big Science, the culture and rhetoric of argument are but a small part of large networks of machines, organizations, and interorganizational linkages. Discursive and textual practices are only one of the resources available for today's science to make forceful claims to authoritative and objective knowledge. In fact, rhetoric is probably not even an especially forceful supporter of objectivity, as decades of disbelieved sociological scientism should make evident. Latour (1988, p. 169) repudiates the textual idealism of rhetoricism strongly: "If all discourse appears to be equivalent, if there seems to be 'language games' and nothing more, then someone has been unconvincing."

The objectivity-as-rhetoric position also has difficulties explaining why some statements are more likely trusted as objective than others, despite the fact that almost all scientific statements are couched in the impersonal and disinterested language of empiricism. Rhetoric may be a part of the differential chances of a statement to become accepted as established knowledge, but cannot by itself explain why so many statements disappear unnoticed into the archives despite their objectivist rhetoric. Rhetoric is, at best, a necessary, not sufficient, condition for objectivity.

OBJECTIVITY AS POWER

The critique of objectivity as power has a long and strange history. It starts with German romanticism and its anti-French and antirationalist love for the peculiar and traditional, moves as *Technikkritik* through the teacher-student ties linking Husserl-Heidegger-Marcuse, becomes a staple of

Critical Theory's critical theory of instrumental Reason and, via Nietzsche, of Foucault's *Biopower*, and is finally picked up by some segments within the new social movements and their intellectual wings, such as radical Green ecology and parts of postmodern academic feminism (see Habermas, 1981/1984, chapter IV). The critical theory of objectivity holds that objectivity is a weapon to exclude marginal voices from the dominant "logocentric" narratives of Western rationalism and empiricism, spearheaded by science (see Hawkesworth, 1992, for a review). In its grand and totalizing aspirations, objectivity suppresses the local whispers of silenced and colonized Others and their lifeworlds. Objectivity sugarcoats a cold and uncaring instrumental reason that has an insatiable appetite for controlling and manipulating Nature and people. Objectivity hides its masculine will to power behind the impartial and neutral facade of truth and progress (Code, 1992, p. 2). In Gergen's (1992, p. 105) words,

> Not only does the discourse of objectivity generate and sustain unwarranted hierarchies of privilege—along with an accompanying array of prejudices, hostilities, and conflicts—but many voices are thereby excluded from full participation in the culture's construction of the good and the real.

There are two versions of this critical argument; one moderate, the other radical. The moderate version criticizes not objectivity as such, but only biases and prejudices that are presented in the disguise of objective facts. This remedial stategy does not aim at displacing objectivity altogether but, rather, at debunking false pretensions to objectivity. Such a "revised" objectivity is, in fact, perfectly compatible with objectivity proper, since the goal is to *rescue* "true" objectivity from claims that do not deserve this label.

Not so for the radical version, which has been put forward most vigorously by some segments of postmodern academic feminism. Here, the very notion and goal of objectivity itself are seen as suspect, as the uncaring and imposing tool of a "disembodied" instrumental and male reason that perpetuates oppression and patriarchy. MacKinnon (1987, p. 50), for example, thinks that "objectivity is the epistemological stance of which objectification is the social process, of which male dominance is the politics, the acted-out social practice." Sprague and Zimmerman (1993, p. 260) agree that "the very form science takes in our culture equates it with power and domination." Cook and Fonow (1990, p. 76) blame objectivity for the "political domination of women through their objectification in research." Lorraine Code (1992, p. 2) assures us that

> it is by now a feminist commonplace that the epistemologies of modernity, in their principled neutrality and detachment, generate an ideology of objectivity that dissociates itself from

emotions and values, while granting no epistemological significance to its own cognitive location.

Some radical critics suggest that objectivity is beyond repair and should be replaced by "standpoint epistemologies." These link ways and modes of knowing to interested social locations and positions. Knowledge is always situated and contextual, part and parcel of the ongoing practices of social groups. Each perspective is, therefore, partial, selective, and incomplete. Standpoint epistemologies seek to make the conditions and circumstances of knowledge distributions visible and accountable, denying that there can ever be a neutral and impartial view from nowhere. Rather, there is a multicultural variety of alternative, sometimes incommensurable, standpoints and perspectives, none of which may claim to be the "only objective one."

Standpoint epistemologies are correct to remind us that all knowledge is part of the world, produced by groups placed in social space and historical time. There is indeed no escaping the worldliness of knowledge to some extraterritorial, transcendental "God's eye" view that was free from any of the empirical constraints and restrictions that actual observers confront. At least since Quine and Kuhn, philosophical naturalism and sociological constructivism have been making great strides in extending empirical science to observers and cognition.

Much more controversial, though, is the suggestion that some standpoints, notably those of marginal, excluded, and oppressed groups, are more privileged than other standpoints in their relationship to truth and objectivity. Haraway (1991, p. 191), for example, prefers "subjugated" standpoints "because they seem to promise more adequate, sustained, objective (!), transforming accounts of the world." Hartsock (1983, p. 159) predicts that in "systems of domination the vision available to the rulers will be partial and perverse," while Harding (1986, p. 26) promises that "by starting research from women's lives, we can arrive at empirically and theoretically more adequate descriptions and explanations—at less partial and distorting ones." McCarl Nielsen (1993, p. 25) argues that "the implication for developing a specifically feminist epistemology is that a woman's perspective (if transformed through consciousness-raising) will lead to more accurate, more complex knowledge."

THE STANDPOINT PARADOX

Making standpoints visible in their effects on knowledge can lead to two possible outcomes. The first is a pronounced and unmitigated relativism that restricts claims to validity to the social and cultural radius of a standpoint and its occupants. This kind of relativism allows for as many legitimate views and valid knowledges as there are standpoints, with no systematic privileges afforded any one standpoint, and no independent and neutral methods for reaching agreements between various standpoints. This strong localism and contextualism implies that basic controversies cannot be resolved rationally because it is always possible to retreat into the trenches of a standpoint, and claim that nobody else can really understand. Paradoxically, strong relativism obliges all arguments and points of view to make themselves *weaker* in admitting to their own narrow limitations:

> Those ideas that are validated as true by African-American women, African-American men, Latina lesbians, Asian-American women, Puerto-Rican men, and other groups (!) with distinctive standpoints, with each group using the epistemological approaches growing from its unique standpoint, thus become the most 'objective' truths. Each group speaks from its own standpoint and shares its own partial, situated knowledge. (P. H. Collins, 1990, p. 236)

As opposed to relativism, the second possible effect of making standpoints visible is not to limit, but to *strengthen* the cognitive authority of the more comprehensive standpoint from which this visibilization can be accomplished. By making standpoints visible and accountable, one can become aware of the selective biases and partial viewpoints they introduce into knowledge. The idea, then, is to transcend a standpoint to arrive at a more complete picture of the world. This picture is more complete now because it includes not only the previous view, but also the relationship between it and the previous standpoint (Nagel, 1986, p. 5). Removing biases and distortions then affords the more reflexive, higher)order, standpoint a less distorted and biased access to truth. These higher levels of awareness are what most standpoint epistemologies strive for; they do not want to relativize but to privilege their own views (Brown, 1994, p. 162).

But then, standpoint epistemologies move back toward traditional objectivity, closer to the very "view from nowhere" they set out to dismantle. Either the standpoint of the oppressed is but one of many possible and justifiable alternatives, in which case it cannot claim any special privileges; or the oppressed and marginal are in a privileged position to overcome bias and partiality, in which case their knowledge will be *superior* to that of the ruling classes. It is superior because it is less partial, less selective, less distorted, and more accurate—in other words, more objective. It is aware of its "situatedness" and "contextuality"—but in this awareness becomes *less* partial and biased.

As a result, radical critics of objectivity restlessly oscillate between the postmodern relativism of pluralistic and diverse standpoints, and the privileged perspective that emerges when standpoints and locations are overcome to permit a more comprehensive and complete, or objective, view (e.g., Calas & Smircich, 1992, p. 244). Mary Hawkesworth (1992, p. 468) captures this paradox:

> Objective inquiry cannot be attained within the preserve of privilege—whether it be the privilege of whites, the middle class, or men. The feminist argument for the inclusion of women and people of color within philosophy and science can be understood not in terms of standpoint theories that suggest that specific individuals or groups have privileged access to truth, *but rather in the demands of objectivity*. (My emphasis)

SOME EMPIRICAL DIFFICULTIES WITH THE CRITICAL THEORY

In addition to these logical and conceptual problems, the radical critical theory of objectivity is also empirically questionable, for several reasons. To begin with, it is prone to confuse certain philosophical pronouncements on scientific method with actual scientific work itself. Sprague and Zimmerman (1993, pp. 256–261), for example, chide "positivism" for its domineering attitude and alleged neglect of the "subjective" and "emotional" components of knowing. Their assumption is that "positivism" is a correct description of actual science, so that criticizing positivism amounts to criticizing scientific work in general.

But, by now, there is widespread agreement that philosophy of science, especially positivism, is a very poor guide to actual science. This result of science studies finds even the approval of many practicing scientists (e.g., Weinberg, 1992/1994, chapter 7). Criticizing Fox Keller's biography of McClintock and her allegedly "feminine feeling for the organism," Richards and Schuster (1989, p. 704) correctly point out that

> method discourse abstracts from and floats above the proper cognitive and social complexity of scientific fields, and so it misses everything that now appears of importance in understanding the dynamics of science.

As a result, the radical critique of objectivity does not measure up well to what is known about actual scientific work. Take the laboratory ethnographies. All the ethnographic evidence about how scientists actually do their research shows that it resembles playful tinkering, practical reasoning, and mundane sense-making much more than controlled and systematic intervention and manipulation (e.g., Knorr-Cetina, 1981; Lynch, 1985; Latour & Woolgar, 1986). This is especially true at the innovative and highly uncertain frontiers, where no rigid mechanics of method exists at all, and where controversies make discoveries very uncertain. Here, work is often playful, experimental, and driven by trials and errors because the world is not yet well understood at all. There may be "planned interventions," but they are notoriously prone to failure and revision. At the frontiers, there is not enough certainty and consensus for instrumental control and logocentric domination.

Observers of research fronts report that work looks surprisingly "interpretive" here (Ziman, 1978/1991, p. 138), and that discovery resembles more artistic creation, not technocratic engineering. Goodman (1978) captures this aesthetic dimension in his metaphor of creative science as a way of "world-making." Innovative science is too uncertain and controversial to have a firm and relentless grip on its victimized objects. At the frontiers, scientific work is often intensely personal and passionate—a far cry from the cold and uncaring machine of instrumental reason (Hilts, 1982, p. 12; Polanyi, 1958/1964, p. 143). Creative scientists follow their "hunches" and "smells" rather than uncaring logic or disinterested method (Thorne, 1994, p. 64f.). Many scientists describe their discoveries as a will to beauty and elegance, not power (e.g., Chandrasekhar, 1987; Weinberg 1992/1994, chapter VI).

Ironically, the critique of objectivity as totalizing power is totalizing itself—it does not have a good eye for variations and differences between various ways of doing science. It assumes, essentialistically, that all science is of one piece, molded into one uniform "logic" or masculine drive for control and technological manipulation (Fox Keller, 1990, p. 54). But some science involves decades-long experimental collaborations between hundreds of coworkers gathered around huge pieces of equipment, while other science is more solitary mental work, such as writing philosophical essays on the essays of other philosophers. What can be gained by forcing all of this into one "logic"? There is simply too much variation in the ways of doing science for all of it to follow some unitary instrumental interest in control. In addition, "women" are too diverse to subscribe to one epistemic framework, unless one defends traditional definitions of "essential femininity" (Longino, 1990, p. 188). In fact, the critics of objectivity repeat the main error of epistemology—there is not one method or logic to science, but many methods and logics, tailored to particular specialties, their means of production, and their social structures.

To be sure, there *are* some areas in science where "control" is more possible and prominent—these are areas of Kuhnian normal science that are comparatively routine and predictable, often with focus on direct technical "applications." Structurally, such areas resemble Weberian bureaucracies. They come closest to instrumental reason. But normal scientific pockets of higher certainty are not even the most visible and prestigious areas. The methodical routines of established science are themselves quickly undermined by perpetual change and innovation. That is, any strongholds of certainty are temporary; any instrumental control is highly precarious and provisional.

In fact, as science develops, it *increases* uncertainty by unfolding and elaborating complexity, and by extending into the infinite in all directions (Gell-Mann, 1994; Dyson, 1988). Science advances—but these advances typically *caution* against all-too-eager attempts at intervention and control. As complexity increases, so does the number of unknowns and uncertainties. Control requires a much more complete and secure knowledge of the world than science-in-the-making can provide. Control requires drastic simplifications, which cannot be delivered by rapidly advancing research frontiers. Their business is novelty, not routine.

OBJECTIVITY AND SOCIAL STATUS

Objectivity, then, is neither logically nor empirically tied to domination and control. This hypothesis resonates rather well with recent social and cultural histories of objectivity. Ted Porter (1995), for example, makes the convincing case, for numerous social fields, that objectivity is actually a sign of *weakness*, not strength. Claims to objectivity and neutrality—often backed up by numbers, statistics, and standards—are most likely raised when bureaucratic decisions or scientific actions need to be explicitly justified because they are under attack. When this happens, "objectivity" is mobilized as a defense: Decisions and actions are justified in terms of impersonal rules and regulations when the discretion of decision makers and "experts" seems no longer trustworthy. Objectivity is invoked to defend credibility in crisis; it is a "democratizing" technology of rules that minimizes the personal discretion and arbitrariness of leaders and elites.

Peter Novick (1988, p. 45ff.) also shows how the "objectivity question" in the American historical profession was tangled up in conflicts over professional autonomy and discretion. Objectivity was most frequently invoked when professional historians had to defend themselves against criticism and competition from amateurs, when it was no longer possible to rely on the "natural" authority of expertise. In this situation, professional historians grounded their work in the detached and impersonal rules of historiographic methodology. Objectification was a strategy of defense and justification in a situation of questioned professional discretion and expertise.

In this sense, the "subjectivity" so cherished by feminist standpoint epistemologies may actually be a much more "elitist" and "domineering" epistemic strategy than objectivity. Supporting this point, Mark Schneider (1993, pp. 7, 83) has documented the resistance of cultural and scientific virtuosos against the objective codification of their skills. Charismatic virtuosos resent and resist objectivity; they rely on tacit and informal discretion and personal judgment to protect their areas of expertise from public inspection and accountability. Once skills are codified and Taylorized by objective methods, they can no longer be mystified as a privileged subjective possession and become, at least in principle, accessible to anyone. Virtuosos emphasize the "subjective dimension" of their respective crafts to distinguish their elite status culture from that of "normal" and routine practitioners. That is, defenses of objectivity indicate *embattled* and scrutinized institutions, not firm strongholds of domination and power. Power is most firm when it appears as grounded in the special skills possessed only by charismatic virtuosos who do not share how they know what they know.

Likewise, the cultural semantics of objectivity is inclusive, not exclusive, as its critics maintain. Objectivity, in its modern seventeenth century sense of removing particular perspectives and contingent constraints on knowledge, seeks to *broaden* the community of knowers to include those not privileged enough to attend the scholastic universities, where knowledge was monopolized by religious elites:

> Aperspectival objectivity became a scientific value when science came to consist in large part of communications that crossed boundaries of nationality, training, and skill. Indeed, the essence of aperspectival objectivity is communicability, narrowing the range of genuine knowledge to coincide with that of public knowledge. (Daston, 1992, p. 600)

Again, it may be the more personal, tacit, subjective, and emotional aspects of knowing that, contrary to feminism, are associated with elitism and social power. For the more personal and emotional one's standpoint becomes, the less it remains subject to public inspection and critical appraisal. *Intuition*, not objectivity, is elitist because it claims that it cannot be known how it knows what it knows; unlike objectivity, intuition cannot be measured, replicated, and accounted for. The will to power enthrones subjectivity and intuition, not objectivity, as above and beyond public control. Was not the sharpest critic of objectivity, Nietzsche, also a relentless enemy of democratic culture? He wasn't much of a feminist, either.

SOME SUGGESTIONS FOR A NEW OBJECTIVITY

I believe we need to rescue objectivity from both orthodox philosophy and standpoint epistemologies. This appears to be the plan of some former radical critics of objectivity as well, many of whom now search for a more defensible, less grandiose kind of objectivity, instead of throwing out the baby with the bathwater, and ending up with relativism and nihilism. In a recent review, Forman (1995) registers just such a general trend toward moderation in critiques of objectivity, motivated by a concern not to abandon general, if not universal, standards of reasonableness altogether. Fox Keller (1995, p. 13) now feels the need to emphasize that "shared measures of competence, some degree of consensus about standards, and some notion of academic freedom are certainly necessary to the viability of intellectual inquiry." Pickering (1991, p. 418) now intends to "recapture an appreciation of the objectivity of science," and the feminist historians Appleby, Hunt, and Jacob (1994, p. 247) have recently scolded standpoint skeptics for being "oblivious to the danger of inventing a new absolutism based on subjectivity and relativism." Natter, Schatzki, and Jones (1995, p. 8), in their introduction to a recent essay collection on a "new" objectivity, realize that

> the fact that the human epistemological condition renders traditional objectivity impossible raises the challenge of how to fashion a successor to the received ideal that is both consistent with this condition and yet able to constitute a meaningful and practical alternative to willfulness and nihilism.

To be sure, this "new" objectivity will be much less than what the philosophers had hoped to find. There seems to be widespread agreement now that objectivity is not, and cannot, be universal and foundational, somehow above and beyond society and time. All knowledge is in and of this world; it cannot be otherwise. Therefore, the question Who is the observer? is the central question for any constructivist philosophy and sociology of science. However, once the old privilege of otherworldly objectivity is denied, it cannot be replaced by a new privilege, that is, that of a particular standpoint. As we shall see, objectivity is actually very fragile and precarious, not cold and menacing. In fact, objectivity shares certain traits with love; *philia* is their common origin, trusting their common process.

If the old absolutist and foundationalist notion of objectivity as timeless and invariant knowledge is no longer acceptable, what kind of objectivity will take its place? While there is widespread agreement on the need to dismiss traditional otherworldly objectivity, its replacement candidates are few and controversial. Surprisingly, most of the scattered suggestions that have been offered so far are rather pedestrian, in some ways taking us back to a precritical stage of epistemology.

Take, for example, Appleby, Hunt, and Jacob's (1995, p. 247ff.) "practical realism." For practical realists, "words result (!) from contact with the world" (p. 248). In making statements, they rely less on philosophers and their rules but on "intuitive wisdom" (p. 248). Despite the inescapable tentativeness and imperfections of knowledge, practical realism aims for "accuracy and completeness" (p. 248). There is an independent reality of objects out there after all, and they are "capable of being 'captured' in the mind by words that point back out toward the thing itself" (p. 250). This happy state of affairs encourages the practical realist "to get out of bed in the morning and head for the archives, because there they can uncover evidence" (p. 251).

Pickering's (1991, 1995) "pragmatic realism" is not much more innovative, despite its being presented in the fashionable idiom of postmodernism. For Pickering, a new objectivity appreciates that "knowledge is somehow disciplined by the otherness with which it engages," and is involved in a "continual struggle with resistant otherness in the material world" (1991, p. 418). Coming dangerously close to anthropocentric Aristotelianism, Pickering (1995, p. 6) invests the world with "material agency," which is to say that "the world is continually doing things."[1] Material agency, we are told, is "temporally emergent," but only "in relation to practice," which is "the work of extending, rather than reproducing, scientific culture—in the sense of building new machines and so on" (1995, p. 14). If I follow Pickering here, he means to say that new machines did not exist before.

Upholding at least part of the core dogma of traditional realism, Pickering (1995, p. 183) believes that this material agency "leaks into and infects our representations of it in a nontrivial and consequential fashion," and that our knowledge is, at least in part, a function of "how the world is" (p. 185). On this fairly solid realist basis,[2] objectivity is whatever manages to survive the "mangle of practice," Pickering's central metaphor.[3] Mangling scientists cope with material agency by modeling their plans and goals into an unknown and uncertain future, and by tuning and retuning their plans and goals when material agency creates trouble for them. This mangling takes place within an existing scientific culture, which "constitutes the surface of emergence for the intentional structure of scientific practice, and such practice consists in the reciprocal tuning of human and material agency, tuning that can itself reconfigure human intentions" (p. 21). Piling metaphor on metaphor, Pickering defines tuning as a "dance of agency":

> The dance of agency, seen asymmetrically from the human end, thus takes the form of a dialectic of resistance and accommodation, where resistance denotes the failure to achieve an intended capture of agency and practice, and accomodation an active human strategy of response to resistance, which can include revisions to goals and intentions as well as to the material form of the machine in question and to the human frame of gestures and social relations that surround it. (p. 22)

As far as I can tell, this means that scientists have goals, that they use machines to accomplish them, that these goals sometimes cannot be achieved, and then have to be changed somehow. The "mangle of practice" appears to be little more than a postmodern restatement of pragmatist trial-and-error methods (the "dialectic of resistance and accomodation").

As a final example, Gieryn (1994, pp. 342–343) and Longino (1990, pp. 62–82) offer a list of criteria that must be met if and when a "new" objectivity is to be possible. For Gieryn, objectivity must tolerate various viewpoints without "preemptive epistemic denunciation," encourage dialogues between opposing approaches without committing category mistakes, reward critical examination of presuppositions instead of dogma, appreciate the complexity of the world, and remain open for revisions of knowledge. This list is hardly objectionable, but does not advance us much further than the classical virtues of epistemological liberalism and individualism, as already endorsed by the empiricism of the analytical tradition, and the critical rationalism of Popper and his followers.

Longino's suggestions for a new objectivity are surprisingly pedestrian as well. Science takes place in a "public space" that encourages mutual criticism through peer review within the shared standards of a more or less egalitarian community that believes in a world independent of perception. Since the institutions of peer review and mutual criticism are "social," objectivity is social as well. Objective knowledge, then,

> is produced by a community (ultimately the community of all scientific practitioners) and transcends the contributions of any individual or even of any subcommunity within the larger community. Once propositions, theses, and hypotheses are developed, what will become scientific knowledge is produced collectively through the clashing and meshing of a variety of points of view. (Longino, 1990, p. 69)

What distinguishes this account from Merton's (1973) Ethos of Science? I do not see much new in the "new" objectivity at all and believe it, in many ways, to be a step back before the postempiricist and constructivist revisions in the philosophy and sociology of science.

In contrast to these approaches, I want to build on Luhmann's theory of self-referential, or autopoietic, social systems.[4] It seems to me that, at present, this theory is the only one to offer a novel notion of objectivity—one that neither reduces objectivity to power or rhetoric, nor goes back to a more traditional, preconstructivist and realist objectivity.

OBJECTIVITY AS MEDIUM

As a medium, objectivity is *not* correspondence realism, not a set of rules or methods, and *not* an extra-worldly view from nowhere. Rather, it is a mode of communication. Following Parsons's theory of symbolically generalized media of exchange, Luhmann argues that such media—among them love, money, power, and truth or objectivity—make the acceptance of communications more likely when the sheer number, selectivity, and contingency of all communications make such acceptance *prima facie* rather unlikely. Media reduce the "improbabilities of communication." Communication becomes improbable when neither copresence nor stratification secure the directness of understanding, or cognitive privileges for those at the top. Communication becomes improbable to the extent that individual minds with unique and special inner experiences cannot be expected to share mental states through mutual empathy and introspection. Communication becomes improbable when writing extends possible audiences so much that there are no longer any guarantees that a communication will actually reach someone. Even if some communication is eventually being picked up, chances are that it will not be understood, and even if it is being understood, it may still be rejected. When it is no longer plausible to rely on some preestablished social harmony, an overarching value consensus, or transcendental states of natural agreement between all civilized and reasonable people, media of communication make sure that communications can nevertheless be coupled to previous and future communications.

This coupling occurs when some communication is used as the premises for subsequent communications. Then, a structure is being built up in which not everything remains equally likely. Media generalize coupling; they decouple coupling from immediately shared experiences, from blind trust in certain charismatic individuals, and from privileged insights that are deemed valid in all possible worlds. Media step in when communicative success can no longer be taken for granted because there are too many possibilities of knowledge and experience, and too many reasons for others not to listen or to say "no."

Media of communication achieve some structure and consistency when all structure and consistency must be accomplished, and can no longer be taken for granted, or anchored in some a priori state of affairs. The coupling of communications remains precarious and reversible, though: Any structure can be undone. Media do not and cannot decide exactly what will be communicated, and how other communications will respond to these communications. Media can only determine in what way communications will be coupled insofar as coupling takes place at all. But it may not, and then there is only noise, no structure.

A medium is not a thing or an instrument; rather, it is a basic mode of communication that determines what kinds of things one can do or say in a given system, such as politics, law science, or love. Media are open and closed at the same time: Communications in a given system must employ one specific medium, but this choice does not yet determine what exactly will be communicated. Objectivity organizes communications on a more abstract level than theories or plans. A medium only decides what *kinds* of communications can take place in a given area, and what *sorts* of experiences count as information there.

Media use "binary codes" to assign all communications to either one of two values. For example, the medium of truth/objectivity signals that scientific communications should not be understood and evaluated as love proposals or legal threats, but as statements about the world that can be true or false, objective or subjective. Again, the medium and its binary code do not decide which statements are true or false, but only make sure that future communications, if and insofar as they occur in and as part of science, refer to previous and current communications as true or false, not as beautiful, just, or powerful. That is, media establish a "basic difference" that remains constant in the coming and going of singular truths and falsities, at least as long as the autopoiesis of the system continues.

Science employs the medium of objectivity or truth. Truth and objectivity do not indicate a privileged access to the world. They do not refer to knowledge that corresponds to some external reality. They do not place knowers in an extramundane position with a view from nowhere. They do not even make sure that knowledge will eventually approximate reality in some asymptotic and teleological process of gradual perfection. Rather, truth and objectivity either *happen* (in which case there *is* science), or do not happen. They do not occur outside of society.[5] They do not indicate states of the world, but states of a social system, science. Truth and objectivity are not somehow above and beyond practice, but symbolize the unity of this very practice in all of its actual diversity. But this coupling remains an internal achievement of the system; it does not bring the states of the system into some sort of isomorphism or correspondence with the states of the world. The world as it is, the thing in itself, remains unknown and unknowable. The theory of media and codes is antirealist and constructivist. It dismisses traditional objectivity, and does not return to a precritical stage in epistemology.

Media and their binary codes are not communications themselves, although one can communicate *about* them on a second level, as I do here. If this second-order communication occurs again in science, as it does here, it must also employ the medium.[6] This means that the medium and its code, unlike the communications they frame, cannot be questioned without questioning and disturbing the system as such. Or, in science objectivity can only be questioned objectively, not subjectively.[7] If this happens just the same, irritations occur that test the limits of the medium. If something is subjective, it cannot be part of established scientific knowledge. If something is objective, the assumption is that others would accept this as well, even if they are not present, and even if they cannot be reached and consulted at the moment.

THE BINDING EFFECT OF OBJECTIVITY

This is the "binding" effect of the medium, or its role in mutually coupling and interlocking communications. Objectivity is not factual consensus, but the factual *presumption* that a factual consensus applies to others as well. Objectivity is empirical; it either mediates communications or not, but at the same time, points beyond the accidents of time and place. In science, communications can and must question whether other communications are, in fact, true and objective, but they cannot dismiss truth and objectivity altogether without irritating or even interrupting the operations of the system as such. If objectivity questions itself ("Is it objective or subjective to distinguish between objective and subjective?"), paradoxes and paralyzing self-refutations occur that may exit science, and lead to other kinds of discourse, such as drama, edification, personal confession, or even silence (Ashmore, 1989).[8]

Objectivity means that one can, at least in principle, decide whether statements are true or false, or one can delay this decision until later, when more is known. If they are false, the code attributes this to its opposite value, subjectivity or error, which rules out, at least for the time being, that false or "merely subjective" statements can serve as the premises for any subsequent statements. Falseness and subjectivity are warning signals at the entrance of blind alleys that do not promise any good results and so can safely be ignored, at least for the moment. In this way, truth and objectivity do not determine what will eventually be the case, but their negative values at least exclude the pursuit of subjectivity and error. But these values can be assigned only by communications, not by the world.

As a rule, the medium and its code interpret false statements as "honest" mistakes, not as wicked deceptions or an insurmountable inability to know better. If statements are true, they are true regardless of kinship or sexual attraction or social standpoint. If they are false, they are false because someone has made a mistake, not because they intend to deceive, or are white and male and cannot know any better. Alternatively, probabilities can be assigned. Whichever, the important assumption is that this decision is not based on love or power but on arguments and evidence. These can usually be ranked according to more or less shared criteria of good versus bad evidence, convincing versus unconvincing arguments, and so on. There *can*, of course, be disagreements about what counts as good and bad evidence, or sound versus flawed reasoning. But the code cannot consider the possibility that truth claims are altogether unrelated to arguments and evidence.

If decisions about truth are changed later, and they always will be, then this is interpreted to mean that *learning* has occurred. One did not know then what one knows now. The medium *cannot* attribute theory changes to accidental shifts in tastes or political balances of power, unless this is itself done with good reasons and evidence. If something other than truth and objectivity is suspected to operate in science, this is cause for alarm among those doing this science, and can only be taken by them as a challenge to do it better, to restore the truth and objectivity. If, say, "interests" or "ideologies" are documented to have an effect on research, this very documentation itself can only happen in the medium of truth ("it is true that this interest has caused this finding"). And if some finding has indeed been caused by some interest, this is cause for *suspicion*, not celebration of the "situatedness" of all knowledge.

This means that the medium and its code interpret, say, the history of science as the progressive elimination of error, and its future as approximation to truth. An outside observer, of course, may question this on another level, as the sociology of science does and must do, but this questioning must then recommend itself as a "more realistic" or "more adequate" alternative. Kuhn (1962/1970), for example, could question the old realist dogma that the history of science moved knowledge closer and closer to truth—but he could do so only by suggesting that his way of doing historiography was "more realistic" than the older Whiggish scientific hagiography. The medium can, without further ado, deal, say, with debunkings of rational reconstructions of science's history, and it can deal with "externalist" histories that emphasize power struggles and historical accidents in the development of science. But objectivity cannot admit the possibility that externalist histories are *themselves* not evaluated on the basis of argument and evidence. If it did, there would be no science. An externalist history can demonstrate that the air-pump's victory over Leviathan was an accident, but it cannot be interpreted and evaluated itself as just another accident.

In other words, the code of objectivity has an ineradicable "blind spot" that follows it around like a shadow. As long as one is doing science, and offering one's communications *as* scientific communications, there is no stepping free of the code. Radical critiques of objectivity "jam" this code; they produce irritations because they argue against argument, or argue that there is no argument, only rhetoric or power. This is what Habermas (1990) calls the "performative contradiction" of postmodernism. The code is the internal humming of science; it is what one would hear if one could listen not just to communications, but to communication itself.

TRUSTING

All media and codes are backed up by trust. In science, one trusts in objectivity. The vast majority of scientific communications are accepted on trust, not on independent confirmation, replication, or open-ended discourse. Trust generalizes the readiness to link one's own communications to those of others. As with truth and objectivity, trust is itself not a communication, though it can be communicated about. When that happens, however, trust may already be eroding.

Trust can coalesce around a person, as in love, around social status, as in early modern gentlemen science (Shapin, 1994), or around procedures, as in modern science (Megill, 1991, pp. 310–311). One trusts that samples have been drawn so as to assure their representativeness, that instruments have been tested and calibrated properly to avoid measurement artifacts, that missing values have been substituted according to established rules, or that good reasons could be given in case that did not happen. One trusts that proper methods for controlling alternative explanations of outcomes were followed, and that a paper was scrutinized by competent experts before being published. Nevertheless, it remains possible to dismiss the review process as untrustworthy and overly political, especially if one's papers are frequently rejected. But such a dismissal shows disappointment in objectivity, and indicates the wish to get rid of power, not objectivity. In addition, one may also trust in reputation, as long as it is distributed by and *within* science, and not by, say, fashion or charity. This trust in procedural objectivity explains the continuing significance of "method" in science, despite all the critical debunkings by ethnographers who have observed—hopefully according to some method—that scientists do not follow any methods in their actual work.

In science, suspending trust becomes an option only when repeated efforts at removing persistent and glaring inconsistencies and anomalies fail (Fuchs & Westervelt, 1996). Most anomalies can be attributed to subjective error or incomplete knowledge. In this way, their resolution can be postponed until sometime in the future. This makes it possible to reconcile the *fact* of disagreement among scientists with the assumption that only one truth and only one objectivity exist. But if all renormalizing explanations for drastic inconsistencies and stunning deviations fail, one may finally be ready, if still very reluctantly, to consider the possibility of fraud.

In fraud trials, trust is *locally* suspended. Fraud busters expect to find motives for, and acts of, deception. Failures of trust in science are typically blamed on individuals' lack of honesty and integrity (Broad & Wade, 1982). Individual scientists betrayed the truth because of selfish interests and ideological partisanship. These individuals can then be sanctioned, while the rest of science remains exempt from distrust. Personalized failures of trust do not routinely extend to the entire apparatus of objectivity. If they did, science would grind to a halt, and could only deal with this one theme of distrust, nothing more. One reason for this is that one can only suspect very few people and their communications of fraud at any particular time. Distrust is very exhausting and impolite, especially among colleagues. Even in fraud trials, one continues to trust most other scientists and their statements, especially those that reveal and rectify a fraud. That is, discoveries of fraud cannot themselves be suspected fraudulent or, if they are, then *this* discovery must not be fraudulent, and so on. That is, one must continue to trust in truth and objectivity. Therefore, distrust can only spread to small corners of the world, leaving the huge rest intact and unquestioned, at least for the time being.

In contrast to fraud, *global* failures of trust in science are very rare. They might happen in true Kuhnian revolutions and, to a lesser extent, in pre- or multiparadigmatic fields with high internal fragmentation into separate schools and camps which believe in incompatible foundational myths and ideologies. In these serious crises, science turns into ideology—what Kuhn (1962/1970) calls "incommensurability" of worldviews, standpoints, and perspectives. In fact, the very language of "standpoints" and "perspectives" already signals systematic and prolonged crises in communication and trust. This difference in scale separates incommensurability from fraud, which is more local and individual.

During profound crises, *anything* the opposite camp suggests, not just certain contributions of individual scientists, may be seen as fundamentally flawed, or even unintelligible. This is when the medium and its code fail to secure the ongoing autopoiesis of science itself, not just the acceptance of individual statements or theories. In this case, it does not really matter whether the camps involved are the ethnos versus rational choicers, mechanical corpuscularians versus scholastic Peripatetics, or phallogocentric white males versus the caring yet oppressed. In radical breakdowns of trust, communication itself is disturbed or even interrupted. The networks of cognitive and social interaction fragment or break up entirely. The separate camps accuse each other of not being able to see what the other can see because they are victims of structural and ideological blinders they cannot begin to comprehend, much less overcome. That is, not only does one's opponent not know, but he is also unable or unwilling to learn. The less they communicate, the more the opponents really do start living in different social and intellectual worlds, as Kuhn suspected. But this cognitive incommensurability is, at least in part, a *result* of network breakdown, not its sole cause, as Kuhn would have it.

Global ideological distrust turns science into politics and intellectual warfare. But this is the exception, not the rule, at least in "mature" sciences that produce some facts. Usually, failures of trust are local, individualized, and then observed as fraud and misconduct.

SOME POSSIBLE OBJECTIONS

Let me deal with some possible objections at this point. Science can consider the impact of interests and social forces, even on scientific knowledge. In fact, this consideration lies at the very heart of any sociology of knowledge. However, insofar as it considers this suggestion as a *scientific* communication, it must again exempt it from suspicions about motives. To communicate that science is power as a scientific communication, one must employ the medium of objectivity, and offer some arguments and evidence for this observation. Coercing others, or buying their support, are not valid options; *pure* and *naked* Macchiavellianism will hurt a scientists's ambition much more than it will benefit same.[9] Science *can* consider the possibility that some of its knowledge is due to social interests and power struggles, but it cannot, at least not at the same time, consider *this* communication as nothing but a strategic move in a power game. Or, if it does, then *this* next communication must suspend suspiciousness and start trusting in arguments and evidence. What is more, the insight that some science is influenced by power and interests can only be interpreted *(by those doing this science, not by their observers)* as a call to remedy this situation and restore objectivity.

Similarly, science can consider the possibility that some of it is sexist and uncaring and reflects a male drive to dominate. If this turns out to be the case, then other science must find data and arguments to rectify this deficit and obtain better or more unbiased insights. In other words, it must learn, and assume that all fellow scientists are willing and able to do so, too. This is, if you wish, the "moral order" of science, or its special mode of solidarity (Rorty, 1991, p. 21–34; Shapin, 1994).

This moral order implies that science *cannot*, for example, consider the possibility that *all* of it, or its entire medium and code, is sexist. For there is simply no place in science from which this observation could be made. If it was made inside of science and circulated as a scientific communication, then it would have to claim that it was not sexist itself. But that would mean that not all of science was sexist, after all. Or, the claim that science is sexist is interpreted as coming from outside, in which case it is perceived on the inside as an ideological attack, not a scientific communication (Gross & Levitt, 1994; Wolpert, 1992; Labinger, 1995). In this case, science will counterattack, and the result is ideological skirmishes, not science. This is exactly the state of affairs in the highly polemical current debates over Science and Technology Studies (Fuchs, 1996). There is no third possibility; the code can only accommodate objective/subjective, or true/false, but not objective/subjective/sexist, or true/false/politically incorrect (Luhmann, 1992, p. 195, 208).

Objectivity cannot, for example, survive the recommendation that women should be "recognized as privileged observers partly because they have developed abilities to understand phenomena through intuition and emotionality" (Sprague & Zimmerman, 1993, p. 259). But which phenomena? All of them? Who decides this, and on what grounds? Objectivity breaks down when it is claimed a priori, for all possible worlds and without any evidence at all, that "gender is not simply one of a number of political statuses that require a critical consciousness, but . . . it is privileged even over class as a basic and inescapable limitation on individual thought" (Loughlin, 1993, p. 8). Is this thought limited as well? If not, how can it escape the inescapable? If so, should not the next move be to remove this limitation? Are all women privileged or just feminists? Who gets to decide this, women or feminists?

CONCLUSION

Instead of summing up, let me point out what I think this argument about objectivity as a mode of trusting communication can and cannot do. First, it proceeds at a very high level of abstraction. It cannot account for cognitive and structural variations between various specialties and disciplines. It is too general and abstract to explain the "content" of scientific knowledge, or scientists' everyday work habits and routines. To do this, microstudies are needed that can be more sensitive to local settings and contexts.

However, unless one believes that there is *no* merit in abstraction and generalization, there is room for a higher-level perspective that compares science to other social systems and organizations, such as markets or hierarchies, and sees "objectivity" as a special medium that generalizes trust in procedural domestications of personal, local, and situational contingencies. In this view, objectivity suspends suspicions about motives, at least more so than markets or hierarchies, which more widely assume selfish motives to deceive. Objectivity, in contrast, localizes deception as fraud; it does not assume that *everyone* has an interest in deception, only a few "bad apples."

At this more abstract level, there is, and there can only be, one world, one objectivity, and one truth. If there is more than one truth, then somone has been partial or made a mistake, which can be corrected by learning. If there is more than one objectivity, they are each not objective *enough*, and must be broadened to accommodate the others. Science cannot celebrate subjetivity as such; even most standpoint epistemologies strive to overcome partiality and bias.

To be sure, an outside observer, say a sociologist of science, can (and must) account for empirical variations in scientific cultures, and explain changes in what social groups of scientists hold to be true and objective. The theory of objectivity as medium does not help much in this sort of investigation. However, when it comes, again, to the truth and objectivity of these investigations themselves, there is, again, only one truth and only one objectivity. Is this not true and objective?

NOTES

1. After following the Parisian actor-network theorists in their radical notion of symmetry between Society and Nature, Pickering (1995, pp. 15, 17) is then quick to add that human beings do have intentions, while material things do not. This, of course, raises the problem of how material agency can be symmetrical to human agency when things do not have intentions. In fact, intentions *constitute* agency.

2. Pickering, of course, is at pains to distinguish his "pragmatic" from "correspondence realism." The former adopts a "performative," not "representational," attitude, and supposes that reality is "interactively stabilized," instead of stable. But insofar as the central credo of realism is that the world selects, at least in part, its own representations, the gap between the two realisms does not seem to be all that dramatic.

3. Pickering himself is not quite content with this metaphor, because "if pressed too hard, the mangle metaphor quickly breaks down" (1995, p. 23, n. 37). The mangle is *not* "shirts in, shirts out"—that would not be constructivist enough. It is also different from mangling in the sense that "my teddy bear was terminally mangled in a traffic accident"—that is too destructive. Pickering also seems unsure what mangles are as domestic technologies; the philosopher Suppe told him that "mangles were devices to speed up the ironing." This conflicts, however, with Pickering's sense of the mangle as the "unpredictable transformations worked upon whatever gets fed into the old-fashioned device of the same name used to squeeze the water out of the washing" (p. 23). In this confusing situation, the best thing to do is "to try to take the metaphor seriously enough but not too seriously."

4. I cannot do justice to this theory here, but Luhmann (1989) is a good intro-

duction. My sources are Luhmann (1979, p. 48–60; 1986, p. 18–33; 1989, p. 36–50; 1990, chapter 4; 1992, pp. 167–361; 1995, chapter 3).

5. They do, of course, occur outside of science. But only science specializes in the systematic construction and deconstruction of truth claims in the special medium of objectivity.

6. This does *not*, of course, mean that one cannot also communicate about science's evilness, its dangers for morality, or its costs. But such communications are not part of science. One can debate the merits and perils of using embryos for research, and one can outlaw such research, but if it occurs anyway, its illegality has no bearing whatever on its truth.

7. This remains, of course, a possibility for, say, law or politics, where science can be critized as illegal or too expensive.

8. Gieryn (1994, p. 324), for example, starts his essay on objectivity by telling part of his life (his)story. It is good enough to know that he "recycles and composts religiously"; few of us can say as much. But I am not sure what this does to his argument. Should it be recycled or composted?

9. This is what Latour (1987) does not realize in his political warfare theory of science.

REFERENCES

Appleby, J., Hunt, L. and Jacob, M. 1994. *Telling the Truth about History*. New York: W. W.Norton.

Ashmore, M. 1989. *The Reflexive Thesis*. Chicago: University of Chicago Press.

Bazerman, C. 1988. *Shaping Written Knowledge: Studies in the Genre and Activity of the Experimental Article in Science*. Madison, WI: University of Wisconsin Press.

Broad, W. J. and Wade, N. 1982. *Betrayers of the Truth* New York: Simon and Schuster.

Brown, R. B. 1994. "Knowledge and Knowing: A Feminist Perspective." *Science Communication* 16:152–165.

Calas, M. B. and Smircich, L. 1992. "Re-Writing Gender into Organizational Theorizing: Directions from Feminist Perspectives." Pp. 227–253 in Michael Reed and Michael Hughes , eds., *Rethinking Organization: New Directions in Organization Theory and Analysis*. London, UK: Sage.

Chandrasekhar, S. 1987. *Truth and Beauty: Aesthetics and Motivation in Science*. Chicago: University of Chicago Press.

Code, L. 1992. "Who Cares? The Poverty of Objectivism for a Moral Epistemology." *Annals of Scholarship* 9:1–17.

Collins, P. H. 1990. *Black Feminist Thought: Knowledge, Consciousness, and the Politics of Empowerment*. Boston, MA: Unwin Hyman.

Cook, J. A. and Fonow, M. M. 1990. "Knowledge and Women's Interests: Issues of Epistemology and Methodology in Feminist Sociological Research." Pp. 69–93 in J. McCarl Nielsen, ed., *Feminist Research Methods*. Boulder, CO: Westview.

Daston, L. 1991a. "Baconian Facts, Academic Civility, and the Prehistory of Objectivity." *Annals of Scholarship* 8:337–363.
———. 1991b. "The Ideal and Reality of the Republic of Letters in the Enlightenment." *Science in Context* 4:367–386.
———. 1992. "Objectivity and the Escape from Perspective." *Social Studies of Science* 22:597–618.
——— and Peter Galison. 1992. "The Image of Objectivity." *Representations* 40:81–128.
Dear, P. 1992. "From Truth to Disinterestedness in the Seventeenth Century." *Social Studies of Science* 22:619–631.
———. 1995. *Discipline and Experience: The Mathematical Way in the Scientific Revolution*. Chicago: University of Chicago Press.
Dyson, F. 1988. *Infinite in all Directions*. New York: Harper & Row.
Forman, P. 1995. "Truth and Objectivity." *Science* 269:565–567, and 707–710.
Fox Keller, E. 1990. "Gender and Science." Pp. 41–57 in J. McCarl Nielsen, ed., *Feminist Research Methods*. Boulder, CO: Westview.
———. 1992. "The Paradox of Scientific Subjectivity." *Annals of Scholarship* 9:135–153.
———. 1995. "Science and Its Critics." *Academy* (September-October):10–15.
Fuchs, S. 1996. "The New Wars of Truth." *Social Science Information* 35:307–326.
——— and Steven Ward. 1994. "What Is Deconstruction, and Where and When Does It Take Place? Making Facts in Science; Building Cases in Law." *American Sociological Review* 59:481–500.
——— and Saundra Westervelt. 1996. "Fraud and Trust in Science." *Perspectives in Biology and Medicine* 39:248–269.
Galison, P. 1985. "Bubble Chambers and the Experimental Workplace." Pp. 309–373 in P. Achinstein and O. Hannaway, eds., *Observation, Experiment, and Hypothesis in Modern Physical Science*. Cambridge, MA: MIT Press.
Gell-Mann, M. 1994. *The Quark and the Jaguar: Adventures in the Simple and the Complex*. New York: W. H. Freeman.
Gergen, K. J. 1992. "The Mechanical Self and the Rhetoric of Objectivity." *Annals of Scholarship* 9:87–109.
Gieryn, T. F. 1994. "Objectivity for These Times." *Perspectives on Science* 2:324–349.
Gilbert, G. N. and Mulkay, M. J. 1984. *Opening Pandora's Box*. Cambridge, UK: Cambridge University Press.
Goodman, N. 1978. *Ways of Worldmaking*. Indianapolis, IN: Hackett.
Gross, P. and Levitt, N. 1994. *Higher Superstition: The Academic Left and Its Quarrels with Science*. Baltimore, MD: Johns Hopkins University Press.
Gusfield, J. 1976. "The Literary Rhetoric of Science." *American Sociological Review* 41:16–34.
Habermas, J. 1981/1984. *The Theory of Communicative Action. Volume One: Reason and the Rationalization of Society*. Boston, MA: Beacon.
———. 1990. *The Philosophical Discourse of Modernity: Twelve Lectures*. Cambridge, MA: MIT Press.
Haraway, D. 1991. *Simians, Cyborgs, and Women*. New York: Routledge.

Harding, S. 1986. *The Science Question in Feminism*. Ithaca, NY: Cornell University Press.

Hartsock, N. 1983. "The Feminist Standpoint: Developing the Ground for a Specifically Feminist Historical Materialism." Pp. 283–310 in S. Harding and M. Hintikka, eds., *Discovering Reality: Feminist Perspectives on Epistemology, Metaphysics, Methodology, and Philosophy of Science*. Dordrecht, The Netherlands: Reidel.

Hawkesworth, M. E. 1992. "From Objectivity to Objectification: Feminist Objections." *Annals of Scholarship* 8:451–477.

Hilts, P. J. 1982. *Scientific Temperaments: Three Lives in Contemporary Science*. New York: Simon and Schuster.

Hunter, A. 1990. "Introduction: Rhetoric in Research, Networks of Knowledge." Pp. 1–12 in A. Hunter, ed., *The Rhetoric of Social Research: Understood and Believed*. New Brunswick, NJ: Rutgers University Press.

Knorr-Cetina, K. 1981. *The Manufacture of Knowledge*. Oxford, UK: Pergamon.

Kuhn, T. S. 1962/1970. *The Structure of Scientific Revolutions*, 2nd ed. Chicago: University of Chicago Press.

Labinger, J. 1995. "Science as Culture: A View from the Petri Dish." *Social Studies of Science* 25:285–306.

Latour, B. 1988. *The Pasteurization of France*. Cambridge, MA: Harvard University Press.

———. and Woolgar, S. 1979/1986. *Laboratory Life: The Construction of Scientific Facts*, 2nd ed., Princeton, NJ: Princeton University Press.

Laudan, L. 1996. *Beyond Positivism and Relativism*. Boulder, CO: Westview.

Longino, H. L. 1990. *Science as Social Knowledge: Values and Objectivity in Scientific Inquiry*. Princeton, NJ: Princeton University Press.

Loughlin, J. 1993. "The Feminist Challenge to Social Studies of Science." Pp. 3–20 in T. Brante, S. Fuller, and W. Lynch, eds., *Controversial Science: From Content to Contention*. Albany, NY: SUNY Press.

Luhmann, N. 1979. *Trust and Power*. New York: Wiley.

———. 1986. *Love as Passion*. Cambridge, MA: Harvard University Press.

———. 1990. *Essays on Self-Reference*. New York: Columbia University Press.

———. 1992. *Die Wissenschaft der Gesellschaft*. Frankfurt/M, Germany: Suhrkamp.

———. 1995. *Social Systems*. Stanford, CA: Stanford University Press.

Lynch, M. 1985. *Art and Artifact in Laboratory Science* London: Routledge.

MacKinnon, C. 1987. *Feminism Unmodified*. Cambridge, MA: Harvard University Press.

McCarl Nielsen, J. 1993. "Introduction." Pp. 1–37 in J. McCarl Nielsen, ed., *Feminist Research Methods*. Boulder, CO: Westview.

Megill, A. 1991. "Introduction: Four Senses of Objectivity." *Annals of Scholarship* 8:301–320.

Merton, R. K. 1973. *The Sociology of Science*. Chicago: University of Chicago Press.

Meyer, J. W. and Rowan, B. 1977. "Institutionalized Organizations: Formal Structure as Myth and Ceremony." *American Journal of Sociology* 83:340–363.

Mulkay, M. 1985. *The Word and the World*. London: Allen & Unwin.

Myers, G. 1993. "The Social Construction of Two Biologists' Articles." Pp. 327–367 in E. Messer-Davidow, D. R. Shumway, and D. J. Sylvan, eds., *Knowledges: Historical and Critical Studies in Disciplinarity*. Charlottesville, VA: University Press of Virginia.

Nagel, T. 1986. *The View from Nowhere*. New York: Oxford University Press.

Natter, W., Schatzki, T., and Jones, III, J. P., 1995. "Contexts of Objectivity." Pp. 1–17 in W. N. Schatzki, T. and Jones, III, J. P. eds., *Objectivity and Its Other*. New York: Guilford.

Novick, P. 1988. *That Noble Dream: The "Objectivity Question" and the American Historical Profession*. Cambridge, UK: Cambridge University Press.

Pickering, A. 1991. "Objectivity and the Mangle of Practice." *Annals of Scholarship* 8:409–425.

———. 1995. *The Mangle of Practice: Time, Agency, and Science*. Chicago: University of Chicago Press.

Polanyi, M. 1958/1964. *Personal Knowledge: Towards a Post-Critical Philosophy*. New York: Harper and Row.

Porter, T. 1995. *Trust in Numbers: The Pursuit of Objectivity in Science and Public Life*. Princeton, NJ: Princeton University Press.

Richards, E. and Schuster, J. 1989. "The Feminine Method as Myth and Accounting Resourse: A Challenge to Gender Studies and Social Studies of Science." *Social Studies of Science* 19:697–720.

Rorty, R. 1979. *Philosophy and the Mirror of Nature* Princeton, NJ: Princeton University Press.

———. 1991. "Solidarity or Objectivity?" Pp. 21–34 in R. Rorty, ed., *Objectivity, Relativism, and Truth*. Cambridge, UK: Cambridge University Press.

Schneider, M. 1993. *Culture and Enchantment*. Chicago: University of Chicago Press.

Shapin, S. 1994. *A Social History of Truth*. Chicago, IL: University of Chicago Press.

———. 1995. "Here and Everywhere: Sociology of Scientific Knowledge." *Annual Review of Sociology* 21:289–321.

Sprague, J. and Zimmerman, M. 1993. "Overcoming Dualisms: Feminist Agenda for Sociological Methodology." Pp. 255–280 in P. England, ed., *Theory on Gender/Feminism on Theory*. New York: Aldine de Gruyter.

Staudenmaier, J. M. 1985. *Technology's Storytellers: Reweaving the Human Fabric*. Cambridge, MA: MIT Press.

Swijtink, Z. 1987. "The Objectification of Observation: Measurement and Statistical Methods in the Nineteenth Century." Pp. 261–85 in Krueger, L., Daston, L., and Heidelberger, M., eds., *The Probabilistic Revolution; Volume 1: Ideas in History*. Cambridge, MA: MIT.

Thorne, K. S. 1994. *Black Holes and Time Warps: Einstein's Outrageous Legacy*. New York: W. W.Norton.

Weinberg, S. 1992/1994. *Dreams of a Final Theory*. New York: Vintage Books.

Whitley, R. 1984. *The Intellectual and Social Organization of the Sciences*. Oxford, UK: Clarendon.

Wolpert, L. 1992. *The Unnatural Science of Science: Why Science Does not Make (Common) Sense*. London: Faber & Faber.

Woolgar, S. (ed). 1988. *Knowledge and Reflexivity: New Frontiers in the Sociology of Knowledge*. London, UK: Sage.
Ziman, J. 1978/1991. *Reliable Knowledge: An Exploration of the Grounds for Belief in Science*. Cambridge, UK: Cambridge University Press.

CHAPTER 9

Science Studies through the Looking Glass

An Intellectual Itinerary

STEVE FULLER

EDITOR'S INTRODUCTION

The problem now is to reestablish working relations between scientists, those who study science and scientists, and the general public. In the final chapter, Steve Fuller investigates the possibilities of reestablishing communication between the scientific and the science studies communities. His own contributions in this respect include arranging an early conference in Durham, U.K., expressly designed to discuss such matters, debating Lewis Wolpert on the BBC, speaking on the Sokal Hoax to several audiences, and in general participating in the Science Wars on both sides of the Atlantic. In this chapter, Fuller takes the interesting position of a critic of both *sides in the Science Wars, while arguing that this is not an opposition between the Two Cultures. He traces the origin of STS to an earlier ambition in Britain to "humanize" scientists—the Edinburgh Unit, for instance, was established exactly for this purpose. Later on, however, the members of this unit and others decided to abandon their original mission for a more prestigious-seeming academic game, wielding the new paradigm of "the sociology of scientific knowledge" instead. But Fuller does not let the scientific community off the hook, either. He is, after all, the author of* Social Epistemology *and the founder of the journal with the same name. This means he is interested in the accountability of scientists and even the possibility of a science "by" the people.*

Fuller evaluates some of the strategies in the Science Wars. For instance, he believes that the Sokal Hoax has been much overrated as a "proof" of anything. The editors should have refused to accept Sokal's claim that he had hoaxed them. He also believes it has been a mistaken strategy among some constructivists to resort to case studies to argue their point. "Counter" case studies can always be found by scientists, as Noretta Koertge's recent A House Built on Sand *clearly shows. He concludes by hoping that the narrowly focused Science Wars will catalyze a broader debate between scientists and STSers about the production and distribution of knowledge.*

THE SCIENCE WARS: A CONFLICT WAITING TO HAPPEN

The ongoing "Science Wars" that pits the amalgam of sociologists and cultural critics of science known as "Science and Technology Studies" (STS) against scientists and their philosophical well-wishers was a verbal collision waiting to happen. At least, I was mentally prepared to join the fray when it publicly erupted. Indeed, some of my fellow STSers even suspect that I have promoted the conflict, much as a hard-line Marxist might wish the plight of the working class to worsen so as to force them to see the inevitability of The Revolution. Readers can judge for themselves the aptness of this analogy in what follows, which focuses on the history of STS through my involvement in the field and increasingly its troubled external affairs. While misunderstanding has been rampant on all sides of this dispute, especially as it has snowballed into the mass media, it cannot simply be reduced to that. A good way to characterize this comedy of errors is that, on the one hand, STS has developed sophisticated tools for analyzing the role of science and technology in society but it has singularly failed to apply those tools consistently to itself and so the field remains deaf to how its claims sound to the people they are talking about (i.e., scientists); while on the other hand, the scientific community still lacks a sophisticated understanding of its place in society but realizes that that place is under threat by the spread of STS-style analyses, if not STSers themselves.

This way of putting matters immediately suggests that we are faced with a version of what the British physicist-turned-novelist, C. P. Snow (1905–1980), originally called the "Two Cultures Problem"—that is, a case in which the training of scientists and humanists prevents each group from appreciating the worldview of the other. Most scientists as well as the media have tended to frame the situation this way, but it is deeply misleading—and not because Snow's forty-year-old concerns lack relevance. On the contrary, they are very relevant, but few of the recent commentators are familiar with the historical trajectory that takes us from those concerns to the present ones. Once we confront this suppressed genealogy, we shall begin to understand some of the curious alignments that have transpired over the course of the Science Wars.[1]

THE PREHISTORY OF THE SCIENCE WARS

If we are indeed witnessing a clash of disciplinary worldviews, why have so few humanists and social scientists rushed to the side of their colleagues who make the natural sciences and technology their objects of study? Relatively soon after the Science Wars began in earnest, it collapsed into a debate over the "intellectual integrity" of STS, and much space was devoted to whether STSers were sufficiently "competent" to study science and technology. Predictably, STSers responded by pointing to the vast difference between their methodologies—ethnographic, hermeneutic, discourse analytic—and the ones more familiar to practicing scientists, which typically involve a much greater degree of control and quantification of the phenomena under study. Yet, strangely absent from this aspect of the dispute have been the voices of humanists and social scientists who have produced research exemplary of these more "qualitative" methods in domains outside science and technology. Where have been the leading anthropologists, sociologists, historians, and literary critics of the English-speaking world while STSers have been at the barricades?

If the Science Wars are a Manichean struggle between "The Arts" and "The Sciences," then all humanists and social scientists should feel under threat. But they do not. In fact, in normal circumstances, they hardly register the existence of STS at all. For example, STS is often—and not unreasonably—portrayed as sociology-driven, yet the best-selling introductory sociology textbook in English lacks a chapter on the sociology of science, despite its self-consciously "global" coverage. Moreover, the book's author, Anthony Giddens, the most prolific and widely cited English-speaking sociologist, has never seriously engaged with the sociology of science, though his ruminations have wandered into just about every other branch of the discipline. What is going on here?

The surprising answer to this question is that the *raison d'être* for STS was provided by natural, not social, scientists. In fact, part of the reason for inventing STS was to counteract the image of the natural sciences that social scientists had foisted on the European consciousness over the previous 150 years in the name of "modernity." Basically, a certain idealized view of the organization of authority in the natural sciences was generalized into a model for the rational governance of something called "society" that coincided with the nation-state. This occurred in the nineteenth century, a period that became disenchanted with religious bases for legitimacy, as clerics continued to justify royal privilege and other "traditional" social practices long after they had outlived their usefulness. The man who gave sociology its name, Auguste Comte (1798–1857), also coined the term "positivism," which he regarded as not merely a philosophical doctrine that stressed the value of logic and observation, but more important, a political program for constituting scientists as a managerial elite that assumed the functions of the Roman Catholic Church.

Comte's influence continues to be felt insofar as most people today believe that if any forms of knowledge have a right to claim universality they are mathematics and the natural sciences. In sociology, this attitude expressed itself in a variety of ways that sheltered these disciplines from sociological investigation until the advent of STS. The founder of the sociology of knowledge, Karl Mannheim (1893–1947), explicitly excluded universal forms of knowledge from his inquiries, except insofar as the character of specific societies corrupted their development or tainted their application. Robert Merton (1910–) founded a sociology of science based largely on the pronouncements of distinguished scientists and philosophers on the nature of science. While no sociologist ever thought that cataloguing the opinions of clerics and theologians would have sufficed for a genuine sociology of religion, Merton proved remarkably influential for his ability to link the insider's account of science with norms and values that have been widely regarded as bulwarks against the excesses of totalitarian regimes. The proverbial Mertonian norms of science—universalism, communism, disinterestedness, and organized skepticism—constitute the self-conscious realization of liberal democracy. Not surprisingly, Merton first presented his sociology of science in 1942 to counter Soviet and Nazi claims that the character of science was determined by the class or race of the people practicing it, and hence scientific validity was directly tied to some form of cultural superiority.[2]

An important reason that sociologists have traditionally wanted to promote the natural sciences as the paradigm of rational social order is that it indirectly lent legitimacy to their own activities, especially by enabling them to speak authoritatively to public issues while remaining autonomous from public control. This indirect sense of legitimacy rested on the attempt of social scientists to approximate natural scientific methods (or more precisely, their idealization). As early as 1843, we find John Stuart Mill (1806–1873) justifying a comparative-historical methodology for the social sciences on the basis of its resemblance to the experimental method in the physical sciences. The more familiar "two cultures" style schism between humanists and scientists emerged only in the last quarter of the nineteenth century, once some social scientists—especially mathematical economists and experimental psychologists—began to argue that natural scientific methods could be directly applied to sociological problems.

Until this eruption of scientism, the social sciences were best seen as having helped normalize the natural sciences in a world that was still unused to associating the sources of worldly power with the manual skills needed to do a laboratory experiment. The metaphor of "midwife" or "underlaborer" is thus not inappropriate to capture the relationship between the social and natural sciences well into the twentieth century. For many social scientists, Thomas Kuhn's *The Structure of Scientific Revolutions* (1962) finally returned the compliment by intimating that the defining features of science were ones that, at least in principle, the social and natural sciences could possess to the same extent. While Kuhn consistently disavowed this reading of his book, nevertheless it must be admitted that the conduct of "normal science" in a "paradigm" does not require the substantial financial and technological investments that provide the most obvious basis for distinguishing the natural and social sciences in today's world. Consequently, the critical impulse triggered by C. Wright Mills's (1916–1962) discussion of the "scientific-military-industrial complex" at around the same time was largely quashed by the end of 1960s, as sociologists reconstituted themselves as a "multiple paradigm science."[3]

In contrast, the roots of STS may be found in distinguished physical scientists early this century who objected to science becoming the secular religion of the modern nation-state. In their day they were typically on the losing side of major scientific disputes (e.g., most of them did not believe in the existence of atoms), but they often supplied telling conceptual objections to the reigning orthodoxy, Newtonian mechanics. Nowadays we tend to forget these scientific contributions—though such scientific revolutionaries as Einstein and Heisenberg did not—and focus instead on their seminal "critical-historical" discussions of the origins of the Scientific Revolution. Ernst Mach (1838–1916) and Pierre Duhem (1861–1916) exemplify the opposing ideological sources for this dissent within the scientific ranks.

Mach was a liberal democrat who feared that science's original Enlightenment mission of enabling individuals to judge rationally for themselves the most efficient means to their chosen ends was being thwarted as science was encouraged to set uniform standards of technical competence for society at large. Duhem was a conservative Catholic who believed that science was inherently incapable of providing a unified understanding of reality and hence required supplementation by a spiritualist metaphysics grounded in an existential commitment. For both Mach and Duhem—and for that matter, later historians who have found the expression meaningful—the "Scientific Revolution" was a mixed blessing, in that as scientific inquiry became autonomous from societal values, the early successes of the mathematical and experimental sciences came to dominate the aims and methods of all fields of inquiry, thereby producing both "pseudo-sciences" of the human (i.e., the social sciences) and the "disenchanted" and "alienated" experience that many people today have of science. The prominence of this experience has increased with each of the twentieth-century's two world wars—which historians sometimes respectively mark as "the chemists' " and "the physicists'" war. The school of philosophers founded by Karl Popper (1902–1994), especially Paul Feyerabend (1924–1994), have continued this sentiment into the present day.[4]

The Cold War era marked the next stage in the development of STS. Here we need to return to C. P. Snow's Two Cultures Problem. When Snow first posed the problem in a 1959 lecture, he was read as arguing that the spiritual goals championed by humanists have been historically superseded by material needs that only science can satisfy. But Snow was really making a much more even-handed point: while scientific skills are singularly necessary for the survival of humanity, scientists lack the moral imagination, especially the facility with alternative futures and the values they represent, that is the strong point of humanistic training. Snow's ideal civil servant would thus be equipped with a humanist's sense of ends and a scientist's sense of means. Indeed, the difference in "cultures" that Snow had in mind was not between a bloodless technocrat and an elitist litterateur. Rather, it was between someone like John Desmond Bernal (1901–1971), the X-ray crystallographer and Marxist historian of science, and someone like George Orwell (1903–1950), the novelist and liberal pamphleteer.

In the late 1940s and early 1950s, BBC Radio frequently held debates between scientists and humanists on the future of Western Civilization. The scientists tended to argue that the path charted by scientific materialism made it inevitable that political problems would be solved by technical means. This, in turn, would remove the volatility caused by the protracted public airing of social problems, the supposed root of Fascism's mass appeal. Voting itself may be atavistic in a rationally organized society, if—as hindsight suggested—only a democracy could have enabled the rise of Hitler. For their part, the humanists were born-again liberals, former Communists who could not tolerate the excesses of the Stalinist regime, even if they were committed in the name of the proletarian revolution. Orwell, in particular, was scandalized by the degree to which the scientists would bend over backward to excuse the Soviets—say, by reversing the meanings of common words to make "freedom" appear to be possible only in a collectivist regime. Orwell was convinced that the unmitigated attraction of scientists to Marxism reflected deep totalitarian tendencies in the scientific mind. He subsequently modeled *1984*'s leading ideologue, O'Brien, on Bernal's on-air pronouncements. As Snow saw it, his fellow scientists erred in trying to model political decisions on scientific problems, where it is normally assumed that nature dictates an exact solution. Marx's philosophy of history obviously facilitated this conflation of science and politics, which only served to promote a public image of science as antidemocratic. If the natural sciences held the keys to human emancipation, then clearly their representatives were projecting the wrong image by leaning so heavily on the idea of nature speaking in one overbearing voice.[5]

Snow's efforts to humanize the scientific mind bore fruit once the British Labour Party came to power in 1963 on a platform fueled by what Labour leader Harold Wilson famously dubbed, "the white heat of technology." Soon thereafter Wilson established several interdisciplinary support teaching programs designed to remind scientists that values cannot be reduced to efficient outcomes but are rather invested in civic traditions that command public respect. The historically most important program has turned out to be the Science Studies Unit at Edinburgh University, which by 1970 had evolved into a full-fledged research unit. The teaching unit was made up primarily of trained scientists who, through acquaintance with the works of Kuhn and Bernal's arch-enemy, the philosophical chemist Michael Polanyi, came to realize that scientists had to be "socialized" and "acculturated" just as much as anyone else. These pioneers included David Edge (a radio astronomer who worked for the BBC and eventually founded the leading STS journal, *Social Studies of Science*), Barry Barnes (a chemist who did a postgraduate course in critical social theory at the newly established Essex University) and David Bloor (a psychologist and mathematician with philosophical proclivities). They constituted the core of the "Strong Programme in the Sociology of Scientific Knowledge."

However, before they became a recognizable research group, the Science Studies Unit registered pedagogical successes in persuading science students to shift their career orientations from basic to applied research, which was a sign that they succeeded in integrating scientific expertise into the larger culture. In retrospect, many of STS's public relations problems can be traced to the Edinburgh School's turn away from this initial success in science education to more autonomously defined (but to be sure, more academically prestigious) research pursuits. It almost seems that as soon as STS discovered the social conditions that enable the production of scientific knowledge, the field's main priority became to use that discovery as a formula for its own self-promotion.[6]

STS'S INDUCTION INTO THE "ACADEMIC LEFT" (AND MY INDUCTION INTO STS)

I was drawn to STS as a history and philosophy of science graduate student in the early 1980s, first at Cambridge and then at Pittsburgh, not long after the publication of the seminal books in the field: Barnes's *Scientific Knowledge and Sociological Theory* (1974), Bloor's *Knowledge and Social Imagery* (1976), Bruno Latour and Steve Woolgar's *Laboratory Life* (1979), and Karin Knorr-Cetina's *The Manufacture of Knowledge* (1980). Of these four, only the first two were ever formally assigned in seminars (in

1980–1981) because of their explicit philosophical defense of relativism. However, the latter two have probably made the biggest impression on me—and probably most others who entered STS in the 1980s and 1990s. Their use of a (mock?) "anthropological" method demonstrated to impressionable minds such as myself the difficulty of seeing the special qualities of science—its rationality, objectivity, reliability, validity—in the scientists' native habitats (laboratories) unless one enters already a believer in those qualities. But where I have always diverged from most STSers is in my belief that the founding texts of STS are not research exemplars to be reproduced ad nauseam. Rather, the appeal to anthropology was a rhetorical tour de force that subverted science by its own hallowed method of controlled empirical observation. In that respect, STS's stance toward science is aptly described as "ironic," in having adopted empiricism only because its subjects did, not because the field was intrinsically committed to an empiricist methodology.[7]

In short, I regarded the original laboratory ethnographies as second moments in a dialectic whose first moment was defined by the empirically unrealistic normative views of science that have steadily gained in currency as Comte acquired more converts both in and out of the ranks of science. But how should society regard the natural sciences once stripped of the Comtean hype and explained in terms used to understand less exalted social practices? Simply continuing to describe "science in action" in ever more numerous settings—STS's default trajectory—fails to address the seriousness of this question, since most people have bought the Comtean hype without ever having witnessed a scientist at work. It is by no means clear that a well-publicized, sociologically "normalized" science would be worthy of the resources and significance currently lavished on it. In any case, defining the normative terms for conducting a post-STS science has been at the center of my own research program of social epistemology over the last ten years. Perhaps because their livelihoods depend on it, scientists are much better "spontaneous social epistemologists" than STSers. In any case, I have found the Science Wars a fertile ground for practicing social epistemology, much to the consternation of my STS colleagues who would wish for a more business-as-usual attitude.[8]

I only gradually learned about the increasing dissatisfaction that scientists themselves had with their vocation, as it has become more intimately involved with the maintenance of the social order. We have already seen that STS owes its origins to scientists interested in checking the "anti-social," often authoritarian, tendencies spawned by the scientific mindset in its wanderings beyond the research site. In this vein, one should also include the physics graduates whose careers started with a stint in the armed forces during World War II. There they experienced firsthand the transformation from what one of their number, Derek de

Solla Price (1922–1983), called "Little Science" to "Big Science." The list of Price's fellow recruits reads like a Who's Who of the prehistory of STS: Thomas Kuhn, Paul Feyerabend, Stephen Toulmin, and John Ziman. Each in their own way sooner or later moved in the direction of STS as an expression of their disenchantment with postwar physics' incorporation into the emerging academic-military-industrial complex. This tendency accelerated in the 1960s, as the peak of the Cold War brought an unprecedented number of students into advanced degree programs in the natural sciences, most of whom were destined for employment in research that contributed to the destruction of people, the environment, or both.

My own attempt to conceptualize this emerging difference between social scientists who wished to piggyback on the authority of the natural sciences and natural scientists who wished to disown that claim to authority was triggered by the remarks of a Spanish graduate student, Juan Ilerbaig, who visited a few American STS departments in the early 1990s and picked up on this schism. In response, I drew a distinction between "High Church" and "Low Church" STS, with the High Churchers following the line of the Edinburgh School in cultivating the disciplinary identity of STS, whereas Low Churchers conceptualized STS primarily as a social movement designed to transform the relationship of scientific work to the rest of society. At the time I was working at Virginia Tech, home to the first doctoral-level STS program in the United States. STS's professional identity was a very live issue for the increasing number of students attracted to the program. Also around this time, the National Science Foundation agreed to provide seed money for several graduate programs in the field. While it was clear that the NSF saw the promotion of STS as a means to a sociopolitically more sophisticated science policy (i.e., beyond relying simply on what the most prestigious scientists think ought to be funded), STS programs tended to be run by people whose main concern was with promoting their vision of STS's academic agenda. This difference in goals has not made for a happy employment situation for those obtaining PhDs from American STS programs.[9]

Nevertheless, my recognition of the problem enabled me to address the leading Low Church society in Washington, in January 1993. There I received, inadvertently, my first exposure to the Science Wars. While perusing the shelves of a local bookstore, I ran across the newly published *Dreams of a Final Theory* by the 1979 Nobel laureate in Physics, Steven Weinberg. The book is a high-minded but thinly veiled defense of the Superconducting Supercollider, which was increasingly under fire for overrunning Congressional spending limits. However, the book's truly distinctive feature is the inclusion of one of the first thorough critiques of STS aimed at the science popularization market. It was clear that Weinberg engaged in this exercise, not to give us free publicity but to immunize the reader against arguments that he regarded as spurious yet increasingly influential in science policy circles. After I alerted an electronic mailgroup about Weinberg's attack, David Edge asked me to write an essay for *Social Studies of Science* reviewing Weinberg's book and a British equivalent by the media-friendly, Professor of the Public Understanding of Science, Lewis Wolpert. The essay appeared in February 1994 as "Can Science Studies Be Spoken in a Civil Tongue?" It was probably the first—though certainly not the last—time that scientists' own opinions of their activities were subject to serious discussion in the pages of *Social Studies of Science*.[10]

Weinberg and Wolpert present an interesting study in both contrasts and similarities in the tactics used to wage the Science Wars. The contrasts stand out most sharply. Whereas Weinberg is a theorist in particle physics, Wolpert is an experimental embryologist. Not surprisingly, each sees science as primarily driven by his specific research orientation. Thus, Weinberg speaks of science aiming for "beautiful theories" whose internal coherence resists all attempts at falsification, while Wolpert points to the falsification of fixed ideas as the only guarantee that one is actually doing science. The one has his theories tested on instruments (i.e., particle accelerators) that have little clear application outside the immediate research context, while the other does his experiments in a medical school where applications are in the forefront of the researcher's mind. Interestingly, Weinberg and Wolpert interpret the anthropological method of STS in quite opposed ways. Weinberg envisages the STSer as harboring a superiority complex in the manner of a nineteenth-century anthropologist looking down at the natives, while Wolpert puts the STSer in the role of the nineteenth-century primitive who disdains what she cannot understand—namely, that scientific reasoning does not conform to commonsense. However, neither scientist recognizes the paradoxical politics of their situation: On the one hand, both claim that science establishes "universal" knowledge, yet their sense of universality does not admit of public accountability. It is science *for* the people but not *by* the people. At this point, it is worth recalling that science did not entail the existence of "scientists" until William Whewell (1794–1866) coined the term in the 1830s to signal that one needed to undergo a university curriculum before being deemed competent to pronounce on mechanical matters.

Despite these substantial differences, Weinberg and Wolpert turned out to be alike on several levels. Both confidently asserted that science is one of Europe's gifts to the world. Both identified the original donor as Thales (624–546 B.C.), the Greek normally credited with launching the Western philosophical tradition by attempting to explain nature in terms of one underlying principle (water). Significantly, both spoke of philosophers and sociologists in the same critical breath, even though the two groups have developed as natural antagonists. The reason is that these nonscientific disciplines would extend the public's eligibility to pass judgment on scientific matters, in the name of either "methodology" (philosophy) or "ethnography" (sociology), but typically at the expense of scientists' personal experience and "creativity." In the words of one recent physics Nobelist, the internecine disputes between philosophers and sociologists "can make a scientist feel like an Algonquin whose hunting grounds are being fought over by two colonial powers."[11] Finally, Weinberg and Wolpert could not countenance a politics of science in which scientists "represent" our knowledge of reality in the same sense as a politician would represent her constituency. Interestingly, they engage in diametrically opposed strategies to protect scientists from this larger sense of social responsibility. On the one hand, Weinberg's book implies that scientists virtually own science, though every so often they catch trespassers who try to extract cultural implications without first having acquired a proper scientific training. (We shall see this strategy at work below in the Sokal Hoax.) On the other hand, Wolpert portrays scientists as modest toilers whose competence does not extend beyond the confines of the laboratory. Here one envisages scientists delivering a fully mapped human genome on the public's doorstep, but then quickly moving on to their next megaproject without involving themselves in the political implications of what they have done.

In early 1994, friends at the University of Virginia alerted me to a seminar being run by the retiring provost, developmental biologist Paul Gross, that systematically studied the writings of a variety of feminists, ecologists, AIDS activists, multiculturalists, as well as mainstream STSers. Soon thereafter a book appeared entirely devoted to critiques of these positions: *Higher Superstition: The Academic Left and Its Quarrels with Science.*[12] Coauthored with Rutgers mathematician Norman Levitt, the book's final chapter suggested for the first time that, the field's pretensions to the contrary, STS might be poorly placed to contribute to a progressive politics, given its refusal to countenance a knowledge base independent of its social origins. If STSers believe that knowledge is no more than what those in authority claim, then how can it serve as a basis for liberating oppressed minorities? The implicit answer—that those minorities constitute themselves as communities bound by traditions of local knowledge—is unrealistic in a world whose local affairs are irretrievably entangled with global ones. Here Gross and Levitt leaned heavily on examples from medicine and ecology, where the failure to adopt a "scientific" perspective was held responsible for untold disasters.

Of course, this line of critique was not new. In many ways, Gross and Levitt had simply reproduced the modernist response to postmodernism in the "Culture Wars" that have been sporadically fought in the humanistic end of American academia, once English translations of the French postmodernist (or, as they were originally known: "poststructuralist") theorists Jacques Lacan, Michel Foucault, and Jacques Derrida started to become available in the 1970s. The novelty of Gross and Levitt lay in their being sufficiently conversant with these trends to assimilate STS to them. On its face, it may not seem such a significant accomplishment, but before Gross and Levitt hit the scene, there had been relatively little interaction between STSers and the postmodernists who, over the previous two decades, were instrumental in deconstructing attributions of "value" in literature and "validity" in literary criticism, using arguments similar to those now being used by STSers against science. The most obvious point of contact is the stress that both the postmodernists and STSers place on the "indeterminacy" or "uncertainty" of interpretation, what is sometimes called the "semiotic" dimension of social life. Nevertheless, STS's relationship with postmodernist cultural criticism remains an uneasy one, given that the latter (at least in the American context) more explicitly promotes a political agenda and challenges the canons of conventional scholarship. To distance themselves from these more radical tendencies, STSers have subsequently tended to stress the autonomy of their activities and the standards of scholarship appropriate to them. As we shall see, this has only made STS more vulnerable to attacks by scientists.[13]

Here it may be worth dwelling on the sense in which STSers are naturally aligned with postmodernism—namely, in their belief in the impossibility of providing a "grand" or "master" narrative for science. Accordingly, there are many stories of many sciences to be told, each partially overlapping with, but also partially contradicting, the others, with no overall discernible pattern. In this way, STSers remain open to alternative traditions of science that are usually suppressed by the grand narratives of progress. However, this postmodern turn is epistemologically and politically progressive only under two conditions: (1) the marginalized traditions are highlighted *at the expense* of the dominant ones and do not simply add to the level of babble in the academy; (2) *everyone*, including the natural scientists, relinquish their claim to grand narratives.[14] Unfortunately, neither condition seems likely to obtain. Talk of progress provides a succinct and convenient way of conveying the ends and norms of science in the soundbite culture of media commentary on science. I personally witnessed the rhetorical power of the grand narrative when facing Lewis Wolpert in a BBC radio debate shortly before the conference I staged at Durham, which is discussed in the next section.

Lacking alternative narratives, STS objections can easily come across publicly as the voice of the "village skeptic" who refuses to accept any responsibility for constituting the future of science. Of course, there are ways around this problem, but they involve thinking on a much larger scale and in a more explicitly rhetorical vein than STSers are inclined to do. I shall simply enumerate them here:

1. Situate the apparent progressiveness of science in a still larger master narrative—such as the rise of capitalism—which casts a more equivocal light on science.
2. Observe that the sequence of great scientists whose achievements would be typically mobilized in support of "progress" would not have agreed amongst themselves—nor with us today—about the ultimate goals of their inquiries in which their achievements would count as stages.
3. Argue that the "universal value" of science resembles that of *democracy*, in that both must be actively preserved because of the ease with which it can be corrupted, often by becoming a victim of its own success (e.g., governments become so popular that they turn authoritarian, sciences become so committed to one line of inquiry that they cannot entertain dissent).
4. Characterize STS as part of the "second phase" of secularization that began 400 years ago with the divestiture of state support for religion. STSers thus assume the role of the Protestant Reformers in relation to those who would follow the Comtean impulse to turn science into a new high priesthood. In that case, the grand narrative of progress is little more than a thinly disguised salvation story.[15]

THE BATTLE OF BRITAIN

It turned out that as Gross and Levitt's critique was breaking on the U.S. national scene, I was preparing to assume a professorial appointment in the United Kingdom, at Durham University. Relevant to the subsequent twists and turns in the Science Wars are subtle differences in academic culture between the US and UK. In particular, British academics have a clearer public identity than their American counterparts. Indeed, American academics tend to lack *any* public presence unless they are seen to represent a traditional "minority" interest, such as Blacks or women. Part of this is due to reasons that any democrat should appreciate, namely, the large number of professional academics in the United States confers relatively low status on their work, unless they belong to a group disadvantaged in the society at large, in which case academia is a vehicle of upward social mobility. In contrast, the boundary separating academic and popular writing is more permeable in the U.K. because its traditionally tighter academic job market has turned journalism into a dumping ground for the overeducated. Consequently, one witnesses differences in *degrees* between British academic and popular writing, whereas in the United States one would see differences in *kind*. The symbol of this point is the presence of the *Times Higher Education Supplement* in most kiosks in the U.K. (To American readers: Imagine what *The Chronicle of Higher Education* would look like, if, like the *THES*, it were owned by Rupert Murdoch.)

In addition, there is a more leveled playing field between the "arts" and the "sciences" in the U.K. than the U.S. Britain combines a tradition of self-made entrepreneurs and eccentric scientific geniuses with a (still) relatively strong commitment to state provision for the disadvantaged. This is not the ideal setting for the public support of Big Science, and hence scientists must actively campaign for it on a regular basis. If British scientists are rhetorically more skillful than their American counterparts (as their more frequent appearances in the media would suggest), then that is because their livelihoods depend on it. Finally, although the United States contains the world's largest number of students enrolled in STS graduate programs, STSers probably occupy a higher proportion of sociology professorships in the U.K. than anywhere else. As a result, STS has a public visibility in the U.K. lacking in the United States. Indeed, there is even a recognizable battleground, known as the "public understanding of science," over which scientists and STSers vie on a more-or-less equal footing.[16]

Indicative of this last point is the threat that British scientists perceived from the publication of the first STS book explicitly aimed at a popular market: Harry Collins and Trevor Pinch's *The Golem*. The book consists of little more than some well-known STS case studies, shorn of their controversial philosophical implications, presented in the spirit of opening the door of the laboratory and letting the reader judge for herself what she sees. The effort was presented as a contribution to the public understanding of science. Scientists were advised that it is in their own interest to drop inflated talk of "rationality," "objectivity," and "truth," and to promote instead the more ordinary image of science as "craft" described in the case studies. In that way, the public would learn to have more reasonable expectations of science, and scientists would not feel a need to promise what they cannot deliver. Despite the authors' intent of providing friendly advice to scientists, their rhetoric seriously backfired, as became evident in Collins's abortive exchange with Lewis Wolpert at the 1994 meeting of the British Association for the Advancement of Science, which garnered enormous publicity soon after my arrival in

Britain. In retrospect, I would say that Collins had misgauged how the idea of science as craft would be taken by scientists. When scientists such as Michael Polanyi alluded to science as a craft, they were envisaging it as an artistocratic leisure activity comparable to a sport or game whose value lay in its intrinsic pursuit, not in some specifiable consequences. What Polanyi most certainly did not intend was that science was like more proletarian forms of labor, such as auto mechanics, which seemed to be the level at which Collins was operating.[17]

On learning of the Collins-Wolpert clash, I sent a letter to the *THES* supporting Collins, suggesting that STS would ironically demystify science much as science itself had demystified theology in the previous century. However, instead of publishing my letter, the *THES* phoned me, saying that so many people had written that they decided to do a feature story on the controversy between scientists and STSers. The idea for a large ecumenical conference then crossed my mind, but I did not act until the story actually appeared, so I could see whether the *THES* could be counted as relatively friendly to the STS cause.

The story, which appeared on 30 September, included a wide range of opinions, from a dismissive Richard ("Selfish Gene") Dawkins to more appreciative assessments by physicist John Ziman and biologist Brian Goodwin. In addition, the *THES*'s own editorial that week undercut a pseudo-serious criticism that Dawkins offered of the sociology of science. He argued that sociologists presupposed the validity of scientific knowledge every time they stepped on board an airplane en route to a conference. Be that as it may, the *THES* retorted, sociologists can probably flourish under a wider range of funding regimes than most scientists, and hence are not nearly as tied down to the fancies of a particular society as the scientists themselves are! After that, I counted the *THES* as an ally, but not all of my sociological colleagues shared my enthusiasm. Collins in particular was understandably concerned that the media had already polarized the issue beyond repair, and that the scientists were all too eager—and perhaps too able!—to assail the sociologists with a barrage of soundbites. Indicative of this polarization was that it would be well over two years before a successor to my conference would be staged, this time organized by physicists in the United States. Of course, the interim was marked by a steady stream of gatherings organized by and for either scientists or STSers.[18]

The Durham conference on "Science's Social Standing" took place on 2–4 December 1994, advertised as the first encounter between scientists and STSers with the explicit purpose of coming to terms with each other. Not surprisingly, the first day resembled the first moments of family therapy: pent-up frustrations giving way to periodic outbursts. However, a convivial dinner that night markedly improved the discourse situation the following day. Nevertheless, as a group, the scientists came off as more confident and coherent than the STSers. Regardless of what STSers think about the empirical ungroundedness of concepts like "truth," "rationality," and "objectivity," they remain rallying points around which scientists can show solidarity, despite the disparateness in their fields of study. This point is easily lost on STSers because of the great store we place on the supposed "disunity" of the sciences. Indeed, a big challenge for us is to understand how scientists can believe that science is both "one" and "many"—in other words, that science strives for unity while pursuing distinct paths of inquiry.

This belief has important practical consequences, which were in ample display during the conference. STSers and their apparent fellow-travelers were continually distancing themselves from one another, demonstrating their disunity with a vengeance: feminists castigated Strong Programmers, macrosociologists and microsociologists crossed swords, and various schools of ethnomethodology interrogated each other's practices. (It is also worth noting that however much private sympathy there may have been for Collins, no one from "our" side—other than his disciple Pinch—ever defended him in print; indeed, some publicly criticized him.) In contrast, among the physicists, chemists, and biologists at the conference, whenever one of them made a claim, you can be sure that another would prop it up from his own distinct angle, sometimes "creatively reinterpreting" what the earlier scientist had said. Often this occurred as acts of collective remembering of the grand narrative of scientific progress.

Interestingly, the scientists were most receptive to STS-like reflections when they were seen to be coming out of the mouth of a scientific icon. The more one could find precedents for revisionist views of science in things said by one of their important predecessors, the more scientists were likely to entertain the idea that change has not always amounted to progress. Consider the incongruity between the formulaic character of most scientific training and the open-ended inquiries associated with scientific research at its best. Once historian Graeme Gooday recounted Thomas Henry Huxley's qualms about standardized lab training in physiology, the Professor of Physics at my university bemoaned the tendency for students in his courses to think that uniform methods must produce uniform results. It was only a short step from that concession to at least a partial acceptance of the idea that objectivity is not a natural outcome of inquiry but must be enforced, often at the expense of other scientific virtues such as creativity and critical judgment. If a history of science could be written that presented STS-style reflections as "always already" part of scientific thought, then practicing scientists might be happy to insert STS into their professional training programs.

The above incident illustrated an important lesson of the conference, namely, that STS's uncompromising sense of "otherness"—our tendency to stress the autonomy of our pursuits from those of the scientists we study—throws up the biggest barrier in any attempt to bridge the Two Cultures. Not surprisingly, then, the sessions that centered on STS's distinctive "case studies" sunk like lead balloons. Harry Collins originally suggested the idea, which sounded fine in principle: take three episodes covered by STS and then ask a historian, philosopher, and practicing scientist to comment on them. The relevant STS texts would be circulated in advance, so that panel members could keep their remarks on point. Michael Lynch offered the additional twist of making Wolpert's account of the history of biology in *The Unnatural Nature of Science* one of the cases, to which an STSer (Lynch) would respond.

Despite Lynch's own masterful diagnosis of the flaws in Wolpert's historical and philosophical reasoning, the weaknesses on our side came through in the sessions centering on the STS case studies. Specifically, what does one say to a scientist who dismisses a sociological account of recent particle physics on the grounds that quarks were eventually accepted by scientists who operated in contexts quite unlike the original one and that therefore the vividness of the STS case study trades on presenting an incomplete account of what happened? In short, case studies can always be accepted "on their own terms" but then ultimately dismissed as lacking clear implications for the grand narrative of scientific progress. Since the postmodernist scruples of most STSers inhibit the development of counternarratives of the history of science, scientists have no problem rhetorically reducing case studies to carping from sidelines, deviations from the norm, or mere plot diversions.[19]

THE SOKAL HOAX

A couple of months after the Durham conference, I was invited by Andrew Ross to contribute to a special issue of *Social Text* on what he had now started to call the "Science Wars." As the leading cultural studies journal in the United States, *Social Text* was keen to adopt a leadership role in bridging the Two Cultures, specifically by demonstrating (leftist) ideological affinities among a wide range of humanists, scientists, and activists that transcended disciplinary differences. The idea was to prevent the controversy sparked by Gross and Levitt from playing into conservative critiques of "political correctness" in higher education that have been a staple of public intellectual life in the United States since Allan Bloom's best-selling *The Closing of the American Mind* (1987). Their avowed leftist credentials notwithstanding, Gross and Levitt were among the keynote speakers at the 1994 annual conference of the National

Association of Scholars, the leading American society for conservative academics. Another conference ("The Flight from Science and Reason") in May 1995 sponsored by the New York Academy of Sciences and specifically based on their book was funded by Bloom's own benefactor, the John Olin Foundation. *Social Text*'s strategy seemed an impeccable one—or so it seemed, until it decided to include an unsolicited piece by Alan Sokal, an obscure theoretical physicist at Ross's own institution, New York University.[20]

This piece, the basis of the notorious "Sokal Hoax," turned out to be the most heavily documented article in the special issue, though the editors of *Social Text* did not know quite what to make of it. Thus, Sokal's article was stuck at the end of the issue and Ross made no reference to it in his otherwise synoptic introduction. On its face, Sokal was trying to derive politically progressive conclusions from the twentieth-century revolutions in mathematics and physics, mainly by a juxtaposition of quotes and commentary that showed that postmodern critics and major scientists reinforced each other's claims. The argument consisted mostly of analogies between the ideas of the scientists and the critics, as well as some rudimentary STS-style explanations that showed the alignment of certain social interests with certain scientific ideas. This style in itself was unexceptional in postmodernist circles and, indeed, was rather clearly presented by current standards.

When refashioning as postmodernism for the French market Daniel Bell's original insights into postindustrial society, Jean-François Lyotard had cited such developments as Einstein's theory of relativity, Heisenberg's uncertainty principle, Goedel's incompleteness theorem, and Mandelbrot's fractal geometry. However, Lyotard had not been as eager as Sokal to implicate a wide range of humanistic scholars in the emergent worldview. Why, then, did Sokal go through the trouble of trawling through the cultural studies and STS literatures, when he could have reached the same conclusions simply by sticking to the writings of his illustrious scientific predecessors? This was the question that occurred to me on first reading the piece, which struck me—and the editors—as a somewhat ingratiating but good faith effort by a natural scientist to bridge the Two Cultures. Nevertheless, Sokal was a tenured professor of physics who declared (sincerely, so he still maintains) that he taught mathematics under the progressive Sandinista regime in Nicaragua in the 1980s. Therefore, given Sokal's clear mastery of the relevant humanistic sources, both his mastery of the scientific sources and his political correctness were taken for granted. Unfortunately, while Sokal did not actually misstate any physical formulae, he managed to include characterizations of texts in his own fields of expertise that any fellow practitioner would have recognized as erroneous. The fact that *Social Text*'s editors did not let a physicist or mathematician vet the article thus turned out to be the grounds on which Sokal declared that he had hoaxed the editors.[21]

The media were more than usually eager to jump on the story. After all, both Sokal and Ross, himself already a media darling, were both in the twilight of their youth (aged 40), located at the same university (which happened to be in New York City), and hired by departments that had witnessed contrasting fortunes in recent years (physics was in decline, while cultural studies was ascendant). Sokal first declared his hoax in the May 1996 issue of *Lingua Franca*, a magazine whose place in college bookstores corresponds to that of the tabloids found at supermarket checkout counters. The story soon thereafter appeared as a front-page story in *The New York Times* and a leading editorial in the *Washington Post*. Newspapers throughout the United States included the story, and the popular right-wing radio and television commentator, Rush Limbaugh, made much of it on his shows. Already something had gone awry. Sokal claimed that his hoax was primarily intended as a wake-up call to postmodernists who were rendering the left impotent by disavowing science. His means was to show that leftist humanists had become derelict in their intellectual responsibilities. Unfortunately, the public uptake of the hoax followed the example of Gross and Levitt's book in playing into the right's call for greater scrutiny of academic practices more generally, a conclusion that Sokal himself had not wished to draw, given that his own work is largely theoretical and hence unlikely to attract the large grants that increasingly pass for "relevance" in market-driven public accounting schemes. Much of Sokal's subsequent explanation of his hoax to the media has made this point, but largely to no avail.[22]

In retrospect, it might be argued that Sokal's point would have come across more clearly had he not tried to expose the scientific illiteracy of STSers and stuck instead to a straight critique of STS's postmodernist political pretensions. But in that case, all that would have been shown is that a physicist can become good enough in science criticism to criticize the critics by their own standards. Since such reflexive critiques are routine in STS and cultural studies more generally, there would have been little that was newsworthy in Sokal's achievement. Indeed, his standing as a physicist might have enhanced STS's credibility as a field whose sense of objectivity transcends disciplinary ideologies. Therefore, the rhetorical power of Sokal's hoax depended on his blurring the boundary between the *controversial* and the *incompetent*. The blurring was made possible by Sokal's appeal to a vulgar realist epistemology that enabled assumptions popularly associated with the conduct of the physical sciences to constitute the standard for evaluating STS work. The result made STS-style critiques of science appear based entirely on bad philosophy. In practice, the strategy worked by providing highly tendentious glosses of what STSers

Figure 1. Translating scientists and STSers

When STSers say . . .	Scientists hear . . .
Science is socially constructed.	Science is whatever enough people think it is.
The validity of scientific claims must be judged from the claimant's perspective.	There is no distinction between reality and how people represent it.
STS is autonomous from science.	STS is disrespectful and willfully ignorant of science
Science is only one of several possible ways of interpreting experience.	Science is merely an interpretation that distorts the true nature of experience.
Gravity is a concept scientists used to explain, say, why we fall down and not up. There are other ways of explaining the same phenomenon.	Gravity exists only in our minds and, if we thought otherwise, we could fall up not down
A scientist's own account of her activities is not necessarily the best explanation for those activities.	A scientist's own account of her activities can be disregarded when explaining those activities.

mean when they make certain counterintuitive claims about their work and its relationship to science. Figure 1 illustrates some relevant moves:[23]

However, as my use of the words "vulgar" and "popularly" suggests, Sokal's own grasp of the philosophy of science is less than completely secure. In the first place, realism is—and has always been—a controversial position within philosophy. There continues to be much debate over whether there is a metaphysically significant distinction between reality and its representations, objects and our concepts of them, and so forth. The tendency to conflate epistemology and ontology is not an intellectual failure but a philosophical standpoint nowadays called "antirealism," the most famous species of which is "verificationism," the house philosophy of logical positivism. Indeed, most of the revolutions in mathematics and physics in the twentieth century, which inspired the logical positivists and in whose lineage Sokal's own theoretical work stands, were born of a rejection of realism and an acceptance of the essentially conventional and instrumental nature of scientific inquiry that is not so far from STS's own philosophical starting point. Thus, it has been both striking and amusing to watch Sokal and his defenders carefully disentangle the postmodernist-sounding utterances of their great predecessors from the nuggets of "real science" that they encase. The best example is Steven Weinberg's highly publicized defense in *The New York Review of Books*, which features this gem: "Heisenberg was one of the great physicists of the twentieth century, but he could not always be counted to think carefully, as shown by his technical mistakes in the German nuclear weapons program." Not surprisingly, neither Sokal nor Weinberg has been treated kindly by the professional historians and philosophers of science who have bothered to comment on the hoax.[24]

No analysis of the Sokal Hoax would be complete without a discussion of Sokal's own public account of his activity as an "experiment." Experimentation has always been a controversial method in the social sciences because of its necessarily deceptive character. Ethical standards for human subjects research have had to tread a fine line between the integrity of individuals and science's collective search for truth. One traditionally important criterion for determining the appropriateness of experimentation is that the knowledge could not have been gained by other means. Sokal's "experiment" on the editors of *Social Text* fails by that criterion, and thus constitutes an unethical use of experimentation.[25] A good way to understand this point is in contrast with another famous experiment on the journal refereeing process.[26] Interestingly, it dealt with the "peer-reviewed" journals that Sokal's experiment was meant to champion. Here journal editors unwittingly evaluated articles they had previously published, except this time the authors were listed as affiliated with much lower-status universities than when their articles were first published. Few of the editors recognized that they had already published these pieces, but still more embarrassing was that many of the articles were now rejected, often with unanimous agreement among the referees. Since this experiment was conducted both by and on experimental psychologists, the experimenters were accused of trying to sabotage the discipline if not science more generally. Nevertheless, the experimenters countered that it was impossible to determine the extent and character of bias in the peer review process without proceeding in such a deceptive manner. Despite the many anecdotes and suspicions about refereeing bias, until the experiment was conducted there had been little systematically gathered evidence on the matter.

In contrast, Sokal's "experiment" aimed to show something that was already amply on display, namely, that many humanists have derived larger cultural significance from contemporary physics without having an insider's knowledge of physics. Since no one had been especially trying to hide this fact, and its reasons were understood within the communities to which it was addressed, there was no scientific end to justify Sokal's deceptive means. Given that *Social Text* and other cultural studies journals had published articles like Sokal's before, was there any serious reason to doubt that they would do so in his case? I doubt it. While Sokal clearly does not believe that political implications can or should be drawn from physics, his "experiment" failed to illuminate this fundamental disagreement. In contrast, at least the books cited earlier by Weinberg, Wolpert, and Gross and Levitt accorded their opponents enough respect to confront their claims directly, albeit less publicly. Finally, for someone who claims for himself a certain level of political awareness, it is hard to believe that Sokal did not anticipate that his hoax would abet the cause of right-wing academia-bashing, as had Gross and Levitt's work, from which he drew inspiration (and even some quotes). From my own standpoint, this would be the only reason for causing a cessation to hostilities in the Science Wars. Otherwise, I believe the Science Warriors have provided a needed reality check for the disciplinary parochialism—the Kuhnian normal scientism—that passes for STSs academic maturity.

WHAT SHOULD HAVE BEEN DONE? WHAT CAN BE DONE?

Most of those who have followed the Sokal Hoax, regardless of their position in the Science Wars, have come to believe that the best response the *Social Text* editors could have made was simply to acknowledge that Sokal had outfoxed them at their own game and pledge that in the future they would call on the relevant experts as referees when articles are submitted that include substantial discussions of specific technical developments in the natural sciences. I disagree. Given the benefit of 20/20 hindsight, I would propose a more daring response, but one that would have been consistent with the very views that Sokal was trying to prove absurd. Accordingly, the first mistake the editors made was to grant Sokal the authority to speak on behalf of his text and thereby accept the verdict that they had indeed been hoaxed. Instead, they should have stuck to the postmodernist tenet of not privileging the author's intention when conferring meaning on a text. Since the author is only one of many possible voices that constrains how the text "speaks," ultimately the text's meaning is determined by the community of readers that the text attracts. It follows that postmodernism recommends a publication policy that simulates ideal market conditions: A journal would treat publication as a libertarian "opportunity" to be heard, an experiment to see if anyone resonates with what it says. If no one ever refers to the piece, then all that would have been lost is the cost of printing the article in the first place; if some do refer to it, then that will have justified its initial publication. Had the editors pursued this tack, then a fruitful discussion about the value of "peer-reviewed" academic publication might have ensued.[27]

Indeed, the editors could have gone further. They could have argued that competence in physics is not necessary for competence in the cultural implications of physics. This is not because STSers lack standards but because the members of a culture need not fully grasp the technical content of scientific ideas that influence them. I have called this the *outside-in* perspective on science.[28] For an STSer it is more important to understand the metaphorical associations that nonscientists—and often the scientists themselves—derive from scientific ideas than to understand the logical structure of scientific theories. Scientists may wish to protest this methodological strategy, but then they court the charge of hypocrisy. After all, practicing physicists are only a fraction of those who contribute to what physics is. The other contributors include engineers, scientists in other disciplines (including the social sciences) who model their own fields on physics, science policy-makers, physics popularizers (including those who see Buddha and the Tao in quantum mechanics), and their New Age readers—not to mention professional physicists such as Steven Hawking and Paul Davies who are probably better known for their metaphysically oriented popular writings.

From a physicist's standpoint, many of these people may have a deficient understanding of the science, but if the physics community were to disown them, then physics would quickly lose its social standing as a science. Why would democratic governments want to invest billions of dollars on activities from which only a few directly benefit? From a sociological standpoint, all of the misbegotten metaphors and half-understandings that Sokal and Weinberg decry in STS are in fact what enables a broad spectrum of people to relate their own experiences and traditions to the rather elite and alien world of physics research. A truly interesting experiment would be for Sokal and Weinberg to turn their efforts at remediation on admiring New Agers, policy makers, and social scientists whose misunderstandings of physics are often as egregious as those of the STS critics. One unintended consequence of their efforts, however, may be to persuade the admirers that their ardor was misplaced, in which case one might expect more physics research projects to meet the dismal fate of the Supercollider.

In the past five years of the Science Wars, there has been a distinct contrast in the development of the two sides. On the one hand, a wider range of scientific voices have been heard, many of them sympathetic to STS concerns and findings. On the other hand, STS has tended to hide behind the image of an autonomous research community. The Sokal Hoax marked a pivotal transition. Before Sokal, STSers would routinely dismiss scientific criticism of their work as misrepresentations. After Sokal, it has become increasingly common for STSers whose intellectual roots go directly back to the Edinburgh School to dismiss as interlopers fellow-travelers in cultural studies and the postmodern humanities. This purification of the STS ranks revolves around the idea that particular case studies of "science in the making" constitute the core of the discipline. The field's characteristic philosophical positions, arguments, and subsequent influence are then claimed to be little more than abstractions (and, in some cases, diversions) from that core body of empirical knowledge. This is a desperate move that not only gets matters exactly backward but also exposes STS to even more needless attacks and declarations of its own irrelevance.[29] By downplaying its challenge to conventional norms of research, STS ironically renders itself most vulnerable. Specifically, the field invites the reanalysis of its cases in conventional terms, according to which a methodological innovation can all too easily be diagnosed as an epistemological error.[30]

This point harkens back to the dialectical origins of the appeal to case studies in STS, namely, to falsify normative theories of science that took empiricism to be the decisive scientific methodology. Beyond that, there has been very little development of case study methodology within STS. A more useful way to understand the appeal to case studies is in terms of their various rhetorical functions. In addition to refuting normative accounts of science, case studies provide a pretext for the participation of multiple perspectives in an episode that might otherwise not be seen. In this respect, the case study may be seen as a vehicle for empowering the politically disadvantaged. Yet, at the same time, a case study creates an intellectual entitlement for the STSer, placing a burden on the potential critic to somehow repeat the work that went into the case study before being seen as lodging a legitimate criticism. This rather proprietary sensibility has been the source of endless friction between STSers and other social scientists. Moreover, because case studies are typically evaluated in terms of their sheer descriptive adequacy (Does it tell a good story?), rather than any larger normative or theoretical context, they can be of potential use to a wide range of users, most notably those who do not share the STSer's personal commitments. In that respect, the case study is well-suited to the opportunism of the postacademic, contract-research culture in which most STS work is done today. In sum, the case study embodies the unresolved tensions implicit in the social role of the STSer: It constitutes the molten, not solid, core to the field.[31]

What exactly would it mean to win the "Science Wars"? Both scientists and STSers vary widely in their opinions on the matter. Some scientists are mainly concerned with stamping out sloppy scholarship in the academy, others with rescuing the left from postmodern decadence. Some STSers are primarily interested in asserting the intellectual integrity of their own pursuits, while others (myself included) wish to dispel the mystifications that surround the pursuit of science. The goals identified for each side are not entirely incompatible, but they do pull in different directions.

In many respects, the Sokal Hoax exemplifies the problem of trying to address intellectual differences and political strategy simultaneously.[32] Sokal justified his parody as a wake-up call to the academic left, but the subsequent debate has become hung up over intellectual standards, which in turn, has opened the door to much generalized bashing of higher education, something no party to the dispute wishes to encourage. Because the debate occurs as an interdisciplinary squabble (science vs. STS), it is easily subject to a "divide and conquer" strategy by policy- and public-opinion makers, whereby academics feel obliged to "clean up their own house," "close ranks," and so forth. However, a more productive debate would realign the parties so that scientists and STSers who wish to protect the academy from the rest of society could stand on one side, while those who wish to use the academy as a vehicle for reforming society could stand on the other—and then resume fighting. In other words, this repositioned debate would not reproduce natural (disciplinary) divisions within the academy but would force academics to seek constituencies outside academia for whom alternative conceptions of the social role of academics could make a difference to their own activities. In that way, the Science Wars, which seem destined to engulf most academics in some way, may become a catalyst for a multiply registered discussion of the production and distribution of knowledge, which is after all, what STS is supposed to be about.[33]

NOTES

1. Snow (1959). A sensitive philosophical analysis of the Two Cultures problem is provided in Sorell (1992), pp. 98–126.

2. Merton (1957) reprints the 1942 article, along with explications and extensions of Mannheim's pioneering efforts in the sociology of knowledge. Editor's remark: (For a detailed reference to Merton's 1942 article having been published under three different names, see chapter 5).

3. On the generally insidious influence of Kuhn on the social sciences, including STS, see Fuller (2000), chapters 5 ff.

4. The account provided in the last three paragraphs is explored in greater depth in Fuller (2000), chapter 7.

5. The best account of the developments sketched in the last two paragraphs is Werskey (1988).

6. For a fuller account of the history, see Fuller (1995b); Edge (1995).

7. Barnes (1974), Bloor (1976), Latour and Woolgar (1979), Knorr-Cetina (1980). As a graduate student, I provided one of the earliest sympathetic philosophical analyses of the sociological perspectives informing STS: Fuller (1984). Of the STS pioneers, Steve Woolgar has been most alive to the ironic character of STS research. See his textbook, Woolgar (1988).

8. Social epistemology has been developed through a series of articles and books, as well as a journal by that name. For a recent elaboration, see Fuller (1993). A textbook version is Fuller (1997). The most comprehensive statement of the disillusionment with science in the Cold War era was Ravetz (1971), many of whose theses would be subsequently picked up and developed by STS, including social epistemology. A crucial presupposition of social epistemology is that the division of intellectual labor into academic disciplines is an accident of history that does more to hinder than to help the production of knowledge. This presupposition turns out to be quite controversial in the context of the Science Wars, since many STSers believe that participation in the dispute should be restricted to the explicit attackers and attacked, as if the matter could be resolved by a gentlemen's agreement among "worthy opponents" without worrying about the implications that understandings of science have for those not directly party to the dispute. I find this a strange response for the simple reason that if science pretends to be "universal" knowledge, then everyone has a stake in the character of its production and distribution. Nevertheless, because he wanted to limit the dispute to such "worthy opponents," Harry Collins ultimately refused to appear at the conference I staged at Durham (discussed below), despite my active solicitation of his support. Finally, in July 1997, Collins staged a Science Wars conference at his home university, which featured his closest allies and the scientists whose swords he had (usually unwittingly) crossed.

9. The professional society most closely associated with High Church STS is the Society for Social Studies of Science (4S), whereas the Low Church is primarily associated with the National Association for Science, Technology and Society (NASTS). See Fuller (1992a,b), Fuller (1993), pp. xii–xiv; and Fuller (1994b). Sujatha Raman is responsible for persuading me that the High/Low Church distinction is best seen in terms of alternative institutionalizations of STS, roughly corresponding to discipline (High) versus movement (Low). In the European STS scene, two textbooks exemplify the High/Low Church distinction: Felt et al. (1995), a German High Church text, and Gonzalez Garcia et al. (1995), a Spanish Low Church text. I was assured that the High/Low Church distinction was a good one when I raised it at a symposium in Copenhagen in October 1992 and Bruno Latour, a cosymposiast and perhaps the most famous High Churcher, admitted ignorance that a Low Church even existed. Since that time I have been mindful of an elementary lesson of the sociology of knowledge, namely, that seemingly similar ideas can be generated from radically different social contexts. In France, STS occupies a place in national policymaking forums that is unknown in the English-speaking world, which is largely due to the relatively close-knit relationship between the French state, industry, and academia. Consequently, it is all too easy to misinterpret the purely descriptive character of French STS work as a radical

critique of science's normative aspirations, when in fact it is designed to provide as much leverage as possible for government agencies to exert pressure on industrial providers of scientific innovation. This critique is developed in more detail in Fuller (2000), chapter 7.

10. Weinberg (1992), Wolpert (1992), and Fuller (1994a). The responses by Weinberg and Wolpert, along with my reply, appeared in the November 1994 issue of *Social Studies of Science*. Another exchange between scientists and STSers occurred soon thereafter, in the May 1995 issue, under the rubric of "Science as Culture: A View from the Petri Dish." My own journal had already published the manifesto of a STS research program aimed at making the acceptability of STS accounts to scientists a necessary condition for their validity: Schmaus, Segerstråle, and Jesseph (1992). I have argued for a similar constraint, but not because I believe that scientists enjoy a distinctive epistemological position, but simply as an extension of the democratic principle that no description can acquire normative force unless it enjoys the consent of the described. See Fuller (1993), p. 32. For an interesting discourse-analytic account of the role of electronic mail-groups in fueling the Science Wars in its early stages (1994–1995), see Herf (1997).

11. The quote appears in Gottfried and Wilson (1997), p. 547.

12. Gross and Levitt (1994).

13. Speaking as someone who entered university in the late 1970s, just as poststructuralism was gaining a foothold in humanities faculties, a striking sign of the gap between the Two Cultures is the fact that so few people, both in and out of STS, have noted the similarities between the Culture and Science Wars. This curious sense of historical amnesia stems from a combination of senior scientists only now catching up with the last twenty years of postmodernism and the most recent generation of STSers becoming familiar with, say, Bruno Latour before having assimilated Foucault, Derrida, and—for that matter—Latour's philosophical mentor, Michel Serres. A good example of this "deja vu all over again" that I have witnessed in my relatively brief career is that a supposed mark of STS radicalism is the tenet that Nature is the product, not the cause, of what scientists decide. (STS students know this as Latour's "Third Rule of Method.") This inversion of the normal relationship between cause and effect, ultimately taken from Nietzsche, is no more than a standard trope in postmodernist deconstructions. For a textbook demonstration, along with a refutation that was widely noted in its day, see Cullen (1982), pp. 86–88; Searle (1983). My own early attempts to apply deconstruction to science after having read the relevant French theorists but not much STS are captured in Fuller (1983).

14. On the idea that postmodernism can thrive only when it is the dominant position, see Fuller (1993), p. 290 ff.

15. I pursued the need for new master narratives for science in my inaugural lecture as Professor of Sociology and Social Policy at Durham (30 November 1995). The secularization of science thesis is developed in more detail in Fuller (1996; 1997, chapter 4; 1999, chapter 6 ff). The impoverishment of STSs historical vision is probably most on display in its response to the creationist challenge to evolutionary theory in U.S. high school biology teaching. When not studiously avoiding the controversy, the orthodox STS attitude seems to be to simply leave the decisions to

local educational authorities. Of course, by taking the local as unproblematic (e.g., one could ask why should a town rather than a state or an entire nation be the relevant locale for educational policy), normative judgments about the nature of knowledge are reduced to matters of political fact: Pontius Pilate would be pleased!

16. Symbolic of this relative equality is that in 1998, the U.K.'s Economic and Social Research Council introduced a pilot program of research into public understanding of science. As the first appointed fellow, I ran a global cyberconference, which is reported in Fuller (1998).

17. Collins and Pinch (1993). The incident between Collins and Wolpert was reported in the 16 September 1994 issue of the *THES*. The original discussion of science as craft appears in Polanyi (1957). Credit for the public comparison of science to auto mechanics belongs to Steve Shapin, which he made in the 14 February 1992 issue of the *THES* ("A Magician's Cloak Cast Off for Clarity"), an adaptation of a piece that appeared in the inaugural volume of the journal, *Public Understanding of Science*.

18. The Durham conference, which attracted eighty participants, was organized through our Centre for the History of the Human Sciences and was sponsored by the *THES*, EASST (the leading European STS society), some leading natural and social science publishers, as well as our departments of Sociology, Psychology, Physics, and Philosophy. The conference received significant media coverage, including a full-length article in the *New Statesman* ("Science Friction," 13 January 1995). The *THES* even decided to cover the electronic correspondence surrounding the conference and declared in its 1994 year-in-review issue that the confrontation between scientists and STSers was among the most important scientific events of that year. The keynote papers of the conference—looking at science's social standing from the perspectives of a philosopher, historian, sociologist, and practicing scientist—appeared in the May 1995 issue of *History of the Human Sciences*. The successor conference, organized by Adrian Melott and Phil Baringer of the University of Kansas Physics Department, occurred in late February 1997 and attracted favorable coverage in *Newsweek* ("The Science Wars," 21 April 1997). One of several constructive features of this conference was that it provided an opportunity for members of a wide range of departments at the host university to explain the nature of their work and how it fits into the larger mission of public enlightenment. This format has been subsequently adopted by other universities in mediating the rift between the Two Cultures.

19. The most sophisticated scientific response to date of an STS case study (again, of the discovery of quarks) focuses on the study's incompleteness. Philosophers of science would here accuse STSers of confusing the contexts of "discovery" and "justification." See Gottfried and Wilson (1997). I attempt to rework the distinction to the advantage of STSers in Fuller (1999), chapter 6.

20. Sokal (1996). Most of the materials associated with the Sokal Hoax have been electronically archived on several web sites. The first was produced by a fan, Jason Walsh, a mathematics graduate student at the University of Washington, Seattle. However, it has been superseded by two others, a largely supportive one in Japan and a largely critical one in the United States:

http://www.math.tohoku.ac.jp/-kuroki/Sokal/index.html; http://members.tripod.com/Science Wars/. The Science Wars were just breaking in Japan in March 1998, when the first delegation of STS researchers from Europe and the United States gave a series of lectures there. I was among them and, in January 1999, I published an article in the leftist periodical, Sekai, designed to neutralize the attack by senior Japanese scientists (trained in the West) on the indigenous STS community. I basically argued that "science" functions as a kind of secular religion in the West in a way that it has not in Asia. This point continues the argument pursued in Fuller (1997), which has now been translated into Japanese by Tadashi Kobayashi.

21. The founding text of postmodernism is Lyotard (1983/1979). It builds on Bell (1973).

22. Sokal revealed his hoax in "A Physicist Experiments with Cultural Studies," *Lingua Franca* (May 1996). Responses by the editors and others to the hoax were published in the following issue as "Mystery Science Theater," *Lingua Franca* (July 1996). By staggering publication of the editors' response to the hoax, *Lingua Franca* undoubtedly helped Sokal garner largely positive media coverage. A notable exception was the story that appeared in *The Chronicle of Higher Education*, which suggested that Sokal suffered from status envy for working at a university where cultural studies was more highly regarded than physics.

23. I originally published this figure in the 28 June 1998 edition of *The (London) Independent on Sunday Review* (i.e., the magazine supplement), in advance of the Sokal-Latour debate mentioned in note 32 below. This should give the reader a sense of the significance attached to the Science Wars in the U.K.

24. The quote is from Weinberg (1996), p. 12, who was soon taken to task by the physicist-turned-historian Norton Wise (in the 3 October 1996 issue) for attempting to bowdlerize the historical record. Wise's public exercise in historical validation subsequently cost him a position at Princeton's Institute for Advanced Studies. Sokal himself was given a hostile reception by one of America's leading philosophers of physics, Arthur Fine, on addressing the philosophical implications of his hoax at the 1996 Philosophy of Science Association meetings. A good anthology on the debates surrounding realism in science is Leplin (1984). A notable exception to the philosophical antipathy to Sokal is a well-placed piece by yet another New York University professor: Boghossian (1996). However, Boghossian managed to incite cross-cultural warfare by arguing that, had the editors of *Social Text* been schooled in Anglo-American rather than French philosophy, they would have seen through Sokal's hoax. This led to a front-page story in *Le Monde* and a flurry of commentary that lasted into the first few weeks of 1997. As it turns out, even in the French-speaking world, Sokal had supporters, including a fellow physicist, Jean Bricmont, who together have written (in French, Sokal and Bricmont 1997) on "the postmodern philosophers' fraudulent science," see Dickson (1997).

25. Stanley Fish first raised the ethical charge against Sokal in an invited commentary for *The New York Times* (21 May 1996) shortly after the hoax had broken as a news item.

26. I am referring to Peters and Ceci (1982), which was conducted on neurophysiological psychology journals. Not only did the study show that reviewers

were influenced by the authors' academic affiliation, but also that the discipline had a very short institutional memory, which gives the lie to the idea that science necessarily builds on its past achievements.

27. This suggestion should be seen in relation to my belief, expressed in a review of *Higher Superstition*, that the appeal to metaphorical resonances between scientific and cultural concepts is not generally a very productive strategy for reforming either the academy or society at large. Indeed, most of the French thinkers who are role-models for this practice are less interested in subverting science than in legitimating themselves through science. Nevertheless, I would much rather criticize these scholars in print than prevent their publication and hence deprive others from possibly deriving some benefit from their work. See Fuller (1995a). For an especially clear justification of this viewpoint, see Fish (1980). One suspects that Sokal himself might have resorted to a libertarian defense of publication, had the editors of *Social Text* decided not to publish his article. He could have then claimed that he had met the standards of the field but that he was discriminated against on grounds that he was a physicist and not a professional cultural studies practitioner. In short, had Sokal been foiled in his attempt to accuse the editors of intellectual sloppiness, he could have then accused them of political correctness. The hoax was thus an experiment incapable of falsifying its hypothesis!

28. Fuller (1997), p. 9; but it is already implicit in Fuller (1993).

29. The charge of irrelevance was especially strong from non-Euro-American contributors to a global cyberconference I ran on public understanding of science. See Fuller (1998). It would seem that there is still life in such "pre-postmodern" concepts as ideology, the distinction between science and technology, and the possibility that some forms of knowledge may be valid in regions other than that from which they originated.

30. While it is true that both Gross and Levitt and Sokal attack mostly members of the academic left whose work lies outside of "core STS," the appeal to case studies has not shielded STS from severe scrutiny. Indeed, one of the more highly cited examples of what these scientists decry is a case study: Latour (1988). Moreover, the scrutiny of cases becomes still more severe in Koertge (1998).

31. Collins (1996) provides a vividly rancorous discussion of the case study method in STS.

32. This point became clear when Sokal debated Bruno Latour at the London School of Economics on 2 July 1998. While Sokal's leftist political heart was in the right place, Latour managed to bury the political challenges facing STS in an apolitical cloud of scholastic virtuosity. The lesson learned on this occasion was that epistemology turns out to provide the best refuge for scoundrels. Perhaps most depressingly, the standing room only audience seemed oblivious to the nuances of debate, preferring instead to turn it into a fairly traditional "arts" versus "sciences" slugfest.

33. The idea that STS should seek nonacademic constituencies (i.e., the Low Church perspective articulated above) is inspired by Robert Wuthnow's account of how intellectual movements—"discourse communities," in his terms—have become full-fledged social movements. See Wuthnow (1989), which traces the

European origins of the (16th century) Protestant Reformation, the (18th century) Enlightenment, and (19th century) Socialism.

REFERENCES

Barnes, B. 1974. *Scientific Knowledge and Sociological Theory*. London: Routledge & Kegan Paul.

Bell, D. 1973. *The Coming of Post-Industrial Society*. New York: Basic Books.

Bloom, A. 1987. *The Closing of the American Mind*. New York: Simon and Schuster.

Bloor, D. 1976. *Knowledge and Social Imagery*. London: Routledge & Kegan Paul.

Boghossian, P. 1996. "What the Sokal Hoax Ought to Teach Us" *Times Literary Supplement*, 13 December.

Collins, H., and Pinch, T. 1993. *The Golem: What Everyone Should Know about Science*. Cambridge, U.K.: Cambridge University Press.

Collins, H. 1996. "Theory Dopes: A Critique of Murphy." *Sociology* 30:367–374.

Cullen, J. 1982. *On Deconstruction*. Ithaca: Cornell University Press.

Dickson, D. 1997. "The Sokal Affair Takes a Transatlantic Turn," *Nature* 385 (30 January).

Edge, D. 1995. "Reinventing the Wheel." Pp. 3–24 in S. Jasanoff et al. (eds.), *Handbook of Science and Technology Studies*. Sage: London.

Felt, U., Nowotny, H., and Tascher, K. 1995. *Wissenschaftsforschung: Eine Einfuehrung* (Science Studies: An Introduction). Frankfurt: Reihe Campus.

Fish, S. 1980. *Is There a Text in This Class*? Baltimore: Johns Hopkins University Press.

Fuller, S. 1983. "French Science (With English Subtitles)," *Philosophy and Literature* 7:1–14.

———. 1984. "The Cognitive Turn in Sociology," *Erkenntnis* 21:439–450.

———. 1992a. "STS as a Social Movement: On the Purpose of Graduate Programs." *Science, Technology & Society Newsletter* (September):1–5.

———. 1992b. "Give STS a Place to Stand and It Will Move the University—and Society." *Science, Technology and Society Newsletter* (December):4–6.

———. 1993. *Philosophy, Rhetoric and the End of Knowledge*. Madison: University of Wisconsin Press.

———. 1994a. "Can Science Studies Be Spoken in a Civil Tongue?" *Social Studies of Science*:143–168.

———. 1994b. "The Reflexive Politics of Constructivism," *History of the Human Sciences* 7:87–94.

———. 1995a. "A Tale of Two Cultures and Other Higher Superstitions." *History of the Human Sciences* 8:115–125.

———. 1995b."On the Motives for the New Sociology of Science." *History of the Human Sciences* 8:117–124.

———. 1996. "Does Science Put an End to History, or History to Science?: Or, Why Being Pro-Science Is Harder Than You Think." Pp. 29–60 in Ross, A. (ed.), *The Science Wars*. Durham NC: Duke University Press.

———. 1997. Science. Milton Keynes and Minneapolis: Open University Press and University of Minnesota Press.

———. "The First Global Cyberconference on Public Understanding of Science."

Public Understanding of Science 7: 329–341.
———. 1999. *The Governance of Science: Ideology and the Future of the Open Society*. Milton Keynes: Open University Press.
———. 2000. *Thomas Kuhn: A Philosophical History for Our Times*. Chicago: University of Chicago Press.
M. I. Gonzalez Garcia, Lopez Cerezo, J. A., Lujan Lopez J. L. 1995. *Ciencia, Tecnologia y Sociedad: Una Introduccion al Estudio Social de la Ciencia y La Tecnologia*. (Science, Technology, and Society: An Introduction to the Social Study of Science and Technology) Madrid: Editorial Tecnos.
Gottfried, K. and Wilson, K. 1997. "Science as a Cultural Construct." *Nature* 386 (10 April).
Gross, P., and Levitt, N. 1994. *Higher Superstition: The Academic Left and Its Quarrels with Science*. Baltimore: Johns Hopkins University Press.
Herf, P. 1997. "The Social Dynamics of an On-Line Scholarly Debate." *The Information Society* 13:329–360.
Knorr-Cetina, K. 1980. *The Manufacture of Knowledge*. Oxford: Pergamon Press.
Koertge, N. (ed.). 1998. *A House Built on Sand: Flaws in the Cultural Studies of Science*. Oxford: Oxford University Press.
Kuhn, T. 1962. *The Structure of Scientific Revolutions*. Chicago: University of Chicago Press. (Second edition 1970).
Latour, B. 1980. "A Relativistic Account of Einstein's Theory of Relativity." *Social Studies of Science* 18:3–44.
———, and Woolgar, S. 1979. *Laboratory Life*. London: Sage.
Leplin, J. (ed.). 1984. *Scientific Realism*. Berkeley: University of California Press.
Lyotard, J. F. 1983/1979. *The Postmodern Condition*. Minneapolis: University of Minnesota Press.
Merton, R. 1957. *Social Theory and Social Structure*, 2nd ed. New York: Free Press.
Peters, D., and Ceci, S. 1982. "Peer-Review Practices of Psychological Journals: The Fate of Published Articles, Submitted Again," *Behavior and Brain Sciences* 5:187–255.
Polanyi, M. 1958. *Personal Knowledge*. Chicago: University of Chicago Press.
Ravetz, J. 1971. *Scientific Knowledge and Its Social Problems*. Oxford: Oxford University Press.
Schmaus, W., Segerstråle, U. and Jesseph, D. 1992. "The Hard Program in the Sociology of Scientific Knowledge: A Manifesto," *Social Epistemology* 6 (3):243–265.
Searle, J. 1983. "The World Turned Upside Down," *New York Review of Books* 30 (16):74–79.
Snow, C. P. 1959. *The Two Cultures and the Scientific Revolution*. Cambridge UK: Cambridge University Press.
Sokal, A. 1996. "Transgressing the Boundaries: Toward a Transformative Hermeneutics of Quantum Gravity," *Social Text* 46/47:217–252.
Sokal, A. and Bricmont, J. 1997. *Impostures Intellectuelles* Paris: Odile Jacob.
Sorell, T. 1992. *Scientism*. London: Routledge.
Weinberg, S. 1992. *Dreams of a Final Theory*. New York: Pantheon.
———. 1996. "Sokal's Hoax." *New York Review of Books* 8 August.

Werskey, G. 1988. *The Visible College*, 2nd ed. London: Verso.
Wolpert, L. 1992. *The Unnatural Nature of Science*. London: Faber & Faber.
Woolgar, S. 1988. *Science: The Very Idea*. London: Methuen.
Wuthnow, R. 1989. *Discourse Communities*. Cambridge MA: Harvard University Press.

Contributors

Henry H. Bauer is Professor of Chemistry and Science Studies at Virginia Polytechnic Institute and State University. He is the author of *Beyond Velichovsky: The History of a Public Controversy* (University of Illinois Press, 1984), *The Enigma of Loch Ness: Making Sense of a Mystery* (University of Illinois Press, 1986), and *Scientific Illiteracy and the Myth of the Scientific Method* (University of Illinois Press, 1992). His forthcoming books deal with the relationship between science and science studies and issues raised in the Science Wars.

Bernard Barber is Professor Emeritus of Sociology at Columbia University. In 1952 he published *Science and the Social Order*, the first comprehensive analytical treatise on the sociology of science. He has also published (with Sullivan, Makarushka, and Lally) *Research on Human Subjects: Problems of Social Control in Medical Experimentation; Social Studies of Science* (a collection of his papers); and various other volumes. His most recent book is *Intellectual Pursuits*, (Rowman and Littlefield, 1998) which discusses the sociology of science and knowledge, and addresses the question of the Two Cultures from Snow to Gross and Levitt and including the "Sokal hoax."

Valéry Cholakov is a research associate in the Department of History at the University of Illinois, Urbana. He holds graduate degrees in science and history. He is the author of the book *The Nobel Prizes*, published 1986 in Moscow, and a number of articles, and book chapters.

Stephan Fuchs is Associate Professor of Sociology at the University of Virginia. He is the author of *The Professional Quest for Truth* (SUNY 1992), and of numerous articles on how to dissolve philosophical puzzles by sociological explanations. He has recently completed the first part of a Trilogy, *Against Essentialism*, which develops an organizational network theory of realism and relativism (University of Chicago Press, in press). The other two parts of the Trilogy investigate the social conditions of rationality, and give a sociological account of consciousness.

Steve Fuller, whose doctorate is in history and philosophy of science, is Professor of Sociology at the University of Warwick, United Kingdom (previously at Durham). He is the founding editor of the journal *Social Epistemology*, and has authored several books and many articles, both in academic and popular periodicals. His recent books are *Science* (Open University Press and University of Minnesota Press 1997), *The Governance of Science: Ideology and the Future of the Open Society* (Open University Press: 1999) and *Thomas Kuhn: A Philosophical History for Our Times* (University of Chicago Press: 2000).

Ullica Segerstråle is Professor of Sociology at Illinois Institute of Technology, (IIT) in Chicago. She holds graduate degrees in organic chemistry, communications, and sociology. Her work addresses "good" and "bad" science, science and values, social theory, and issues on the interface between the social and life sciences. She is the author of *Defenders of the Truth: The Battle for Science in the Sociobiology Debate and Beyond* (Oxford University Press (U.K.) (2000) and the editor of *Nonverbal Communication: Where Nature Meets Culture* (Lawrence Erlbaum Associates, 1997; with Peter Molnar). She is a coauthor of "The Hard Program in the Sociology of Scientific Knowledge: A Manifesto" with Warren Schmaus and Douglas Jesseph (1992) and the author of a number of other articles and book chapters.

John Ziman is Emeritus Professor of Physics at the University of Bristol and a Fellow of the Royal Society. After early retirement from Bristol in 1982, he was a Visiting Professor at Imperial College, London. He was the founding Director of the Science Policy Support Group from 1986 to 1991 and the Chairman of the Council for Science and Society from 1976 to 1990. Since 1994 he has been the Convenor of the Epistemology Group, which studies the evolution of knowledge and invention. He is the author of a number of books on solid state physics, and on science, technology, and society, including *Prometheus Bound: Science in a Dynamic Steady State* (Cambridge University Press, 1994), *Of One Mind: The Collectivization of Science* (American Institute of Physics, 1995), *Real Science* (Cambridge University Press, 2000), and *Technological Innovation as an Evolutionary Process* (Cambridge University Press, 2000).

Name Index

Subject Index

T